SEA, SUN & TARAIPO

Millionaires in Time

John Jameson

Grosvenor House
Publishing Limited

This book is published by
Grosvenor House Publishing Ltd
Link House
140 The Broadway, Tolworth, Surrey, KT6 7HT.
www.grosvenorhousepublishing.co.uk

A CIP record for this book
is available from the British Library

ISBN 978-1-83975-124-0

To my grandchildren, Rosie, Maja, Charlie and Zachary.

The world is yours.

For the lucky winner!

'Juggler'

Plymouth August 2020

Preface

This is a story of adventure, of sailing away with no destination, and no plan or time constraints. It started as a dream of sailing on warm seas, and of finding like-minded friends with whom to share the life. I became totally addicted to the rolling oceans and living the life of a sailing nomad.

I was encouraged to write a book about my adventures, by the audiences that came to our lecture tour in France, Scotland, England, the USA, Spain and Switzerland. It was often the non-sailors who were intrigued to learn more about how we handled the loneliness, the danger, the storms and the hardship of living on a small boat.

If you want the complete experience, I recommend that you fire up the map.openseamap.org link on your phone and virtually navigate with me into all of the amazing harbours and atolls into which I ventured.

Acknowledgements

I have to thank many knowledgeable friends and acquaintances for their input and encouragement over the long haul in producing this book. Nicole Kilchoer for her diary, and for keeping the adventure going. Martin Landolt for regular reviews in the pub. Vicky Jackson for the early redirection. Brian Jameson, my cousin, for a massive first edit, plus Alan Gay, Michael Davenport and Catherine Simpson, of Tyne and Esk Writers, and Renate Bishop for the watercolour map.

I should not forget the many cruisers who generously gave me a helping hand along the way.

Last, but not least, my deepest gratitude to all of my fantastic crews who shared the ups and downs of Taraipo's voyage around the world.

Contents

Part One: Escape from the city

When spring comes around and the sap is rising, a schoolboy has the urge to discover pastures new. The distant horizon and a boat to sail there was my fantasy as I lay in the sun at the Didsbury Lido in Manchester.

It was late spring; time to be swatting for exams. I couldn't stay indoors any longer. I got on my bike and went hunting for Hollingworth Lake in the bowl of hills above Rochdale about 20 miles away from our terraced house in Salford. I had heard that there was a sailing club up there. The notice on the entrance gate said 'Private Members Only'; clearly a shy, middle-class 15-year-old doesn't dare to go into an upper-class enclave, full of expensive cars and fascinating high-tech sailing dinghies. I bought myself an ice cream and, weakly, could only watch the dinghies launch and race around the lake. A few weeks later, I returned to watch the sailors tinkering with their boats. I plucked up the courage to go through the gate and ask if it would be possible to crew for somebody. Surprise, surprise, I was welcomed by the Honorary Secretary, who took me out to race his 12' Firefly dinghy. Afterwards, he let me sail the dinghy from the outer jetty back into the harbour. I thought that I knew how to sail and cast off with confidence, disappeared into the distance, capsized and had to be towed back by the rescue launch ignominiously. Clearly, I did not know how to sail into the wind, which requires a zigzag course catching the wind on one side, then the other of the sails. The following weekend and by the strangest of coincidences, I was matched up with Arthur Bullough, a surgeon, who, it transpired, was my mum's boss. Mum was the men's surgical ward Sister. Mr Bullough drove a Humber Super Snipe car, with a picnic hamper and a bottle of red wine in the boot. I had a great summer being chauffeured in style to regattas with his Merlin Rocket sailing dinghy in tow. We didn't win many races; I was not heavy enough to balance the dinghy in strong winds and didn't have the knowledge or experience to help Arthur up his game, but I was keen and definitely had the sailing bug.

Running away to sea

It was no surprise that, after one year at Manchester College of Art, I ran away to sea. College life was great. I shared a loft in town, had two great love affairs, lost my virginity and played double bass in Jim Percival's Davenport Jazz Band. The bass was transported on the rear platform of double-decker buses, or Jim, the band leader, would come and collect me from college. We strapped the bass on the car roof and drove to gigs in pubs and clubs all over the North West, Lancashire, Cheshire and Yorkshire. We even played at The Cavern in Liverpool. On the road one night, the back strap came undone and the bass swung down the side of the car. Jim's reflex was to slow down, reducing the slipstream that was holding my precious bass off the ground. We all yelled and he sped up while we wound down the window and hung onto the bass until the car came to a halt. The 'trad' band was 'semi-pro'; the trumpet, clarinet, trombone, banjo and drummer all had daytime jobs. We were paid a few pounds per night and were given as much beer as we could drink, while still playing in tune or so it seemed to us.

I loved college life, but my tutors didn't love my art. They recommended that 'industrial design' rather than 'wood sculpture' would be the best course for me to take. As that did not sound the fun career I envisaged, I was obliged to leave.

Buxted Chickens was my first employer, delivering frozen chickens to butchers in Greater Manchester. At each stop, I had to stick a poster in the shop window. The job only lasted a week; it kept me in food and cigarettes. My pride did not allow me to return home.

In at the deep end at 19 years old, a telegram arrived at our artist's studio from yacht delivery skipper Alex Brims, RN Rtd (an ex- Naval Air Service navigator) saying 'join ship Lochinvar at Greenock on Friday for a delivery from Clyde to the Thames'. I had not expected such an immediate offer of a job. I had embroidered my sailing experience in the letter of application; in fact, I had never been to sea on a yacht, and I did not own any waterproofs or boots. The next day, I caught the '66' bus into Manchester to the Army & Navy store at Shudehill to see what second-hand oilskins I could afford with the money a failed student did not have. I had to borrow from my mum, not daring to ask my dad. He would not approve of my leaving college, a career in art or running away to sea, although he never said so much to me. His ambition for me was to become a lawyer. All that I could afford to buy was an ex-submariner's stiff, black, smelly oilskin smock, for five pounds. Off I went by train with a one-way ticket, arriving in Greenock on a wet Tuesday evening to discover that the MV (motor vessel) Lochinvar was not a yacht. She was a 210-foot long registered passenger ship, the 50-year old Oban to Mull ferry. Unfortunately there was a writ fastened to the mast by the Harbour Master forbidding her to go to sea as none of the delivery crew members were registered seamen. We were obliged to take along with us three crew from the Seaman's Pool. Their declared interest was to return to Greenock within seven days, in order to claim the dole unemployment benefit. We set sail that night and I did the midnight until three watch, standing alone at the wheel and steering for the Ailsa Craig, a distant lighthouse, with a bucket by my side in case I was seasick. That is how I learnt about the *International Regulations for the Prevention of Collisions at Sea,* as I had to interpret the white, red and green lights of passing ships. Our first port of call was to enter the River Liffey in Dublin in order to bunker with cheap fuel oil. Having a local pilot aboard is obligatory. Not having the appropriate radio equipment, we had to 'make smoke' from the funnel and hoist the 'I require a pilot' flag in order to attract a pilot to guide us into the docks. We rigged a rope ladder over the side to enable him to climb aboard from the pilot cutter.

Underway again, it was my task to make the porridge, in the galley on the upper deck. In the early morning, we were rolling up the English Channel with half a gale blowing and a big swell running when I heard a loud clang behind me, but did not dare to look away from the pan and burn the porridge. Within seconds, I was up to my waste in foam; the fire extinguisher, which had been trapped behind the iron half-door, had fallen over and exploded. I managed to save the porridge and serve it up to the captain on the bridge, for which I was complimented. There was a bit of an embarrassing moment at the end of the voyage as we attempted to bring the ship alongside Southend pier short-handed at night (the professional seamen had deserted us at Littlehampton in order to fly back to Glasgow in time to collect their dole money). Before we had time to tie the stern to the pier, the current had swung it away and she became 'tide girt' at right angles to the pier, so we had to depart and make another attempt. It was all very embarrassing and under the steely gaze of the new owner, the pier-end fair boss, who proceeded to pay us off from a wedge of £100 notes in his back pocket. As a souvenir of my first delivery, I removed a small brass plaque from the '*3rd Class Ladies Toilet*' door.

I was invalided out of the yacht delivery business, sadly missing a transatlantic trip with the famous Americas Cup racing yacht Sceptre. On a delivery from Burnham to Glasgow, through the Forth and Clyde Canal I finished up in Falkirk hospital with osteomyelitis of the hip joint. The cure was to lie in bed with one leg under traction, and frequent injections of antibiotics. There were fun moments later, chasing nurses with my wheelchair. I would approach from behind, catch them below the knees and have them land in my lap! The girls got their revenge. I woke up in the middle of the night to find myself pinned down while they shaved off just one half of my beard. I had enough cash in hand to afford a newspaper each day and buy 'Senior Service' cigarettes, which we were allowed to smoke in the conservatory at the end of the long ward. There was no connect with the guy in the bed next to me, a heavy smoker coughing up his lungs! After a long stay, and now on crutches, I was sent home by train accompanied by a nurse. Those were the days when the NHS gave a bespoke service.

The next summer job was teaching sailing at the Barge Hound Sailing School in Wooton Creek on the Isle of Wight. Staff and students lived aboard the barge. Hound floated on the high tide twice every 24 hours and sat on the mud at low tide. She leaked a bit and it was the Chief Instructor's responsibility to remove the bung at low tide to drain the hull. I dreamt I saw my wellington boots floating past my bunk. The next morning, there was a muddy tide mark on the cabin sides; the chief had forgotten to replace the bung – my boots had been afloat! My schoolboy band *The Blue Five* came for a week. I had managed to get a gig for us at a floating nightclub on the barge opposite. The boys covered themselves in glory by turning the 'fruit machine' upside down and emptying out all the sixpences. The club owner was glad to see them go back to Salford.

I experienced my first dismasting with a sailing school dinghy in the middle of the Solent. One minute, the sails were full of wind, then suddenly all was quiet, and the mast and sails were in the water. The problem then was disentangling the mess and getting a tow back to shore. All good fun and a steep learning curve for JJ the skipper, and the crew, and an expensive afternoon for the owner of the Barge Hound Sailing School.

In between sailing jobs, I managed to find myself navvying on Sidmouth beach. We were laying a sewer pipe into the sea, a challenging project for the young engineers and a bunch of navvies with no experience of the sea. On day one in a helmet, I lost my job as a brickie's mate. I was wheeling a barrow full of cement along a plank, couldn't hold it, lost my balance and fell six feet to the ground. Only my pride was hurt! I scraped up as much of the cement as possible and the bricklayer took over. I was relegated to the noisy jackhammer, digging a trench for the pipe all day long with no ear protection. To lay the pipeline, the engineers at head office had designed a gantry with an air tank on each corner. It would straddle the trench into which the 1.5m pipe was to be laid. This contraption was then to be pulled down the pebble beach and along the seabed with a Tirfor wire rope winch attached to a 1-ton Danforth anchor. The company had bought a wartime American DUKW six-wheel drive amphibious truck to carry the anchor into the sea. The anchor was craned onto the back of the DUKW and tied on with a quick release knot. Sadly, on releasing the knot, a fluke caught the side of the DUKW as the anchor slid off the back; the DUKW filled with water and sank. This was the moment when the driver realised he didn't know how to swim. His mate kept him afloat. The company had bought some Marksway lifejackets – the air inside squeezed out if you fell in head first! During the first attempt to pull the gantry into the sea, the winchman was seriously injured when the wire snapped and whipped around his body. Men were an expendable commodity in those days.

At night, the Conservative club was the place to mingle. It was summer, lots of tourists around and I was flush. For the first time in my life, I was paid cash, in a brown paper envelope every Friday afternoon. My new girlfriend's father, a local builder, poured his earnings into the club's 'one-armed bandit'. He calculated that he broke even with the machine over a 12-month period. Sad. By contrast, Cynth, his daughter, an art student, was vivacious, auburn, big earrings and enjoyed a pint. Sadly, she disappeared off my radar when she went back to college and my job on the beach ended. Her little sister, Gloria, would occasionally turn up at my bachelor pad for a cuppa; otherwise, I spent a lonely winter commuting to Exmouth to work in Lavis's boatyard. The work was interesting. We built a traditional wooden yacht from the keel up, with steamed planking. You had to look out for yourself or get injured. Ginger, one of the lads, passed out while being pushed, arms forward holding tin and brush, into the narrow end of a racing yacht to varnish it. Three of us were nearly crushed when a five tonner we were holding upright started to fall on us. The last boy to let go was caught by a falling prop, but escaped with minor injuries.

Spring came and I was glad get back afloat at the Island Cruising Club in Salcombe, South Devon. Sailing instructors didn't get paid much, enough for cigarettes and beer, but we had a wonderful lifestyle. We all lived together (50 staff and members) aboard a 120' Fairmile Motor Torpedo Boat, and went sailing every day in dinghies and keelboats. I was promoted to 'John the Launch', one of the best jobs I ever had, ferrying staff and sailors six days a week from the 8am sleepy office girls run ashore until the return of the club's barmaid at 11pm. Lots of flirtations and all good, clean fun.

In the winter, we moved ashore and shared flats around the town. We did maintenance work on the boats and had lunch together in Charlie Yabsley's restaurant. The last person out of the boat shed at lunchtime shut down the 'Terrapin' diesel space heater and topped up the fuel tank. It was a crude heater consisting of a 2m chimney atop the fuel tank. The first person back after lunch fired her up, and away she roared. The tank had been topped up in error with petrol! Imagine the potential for a mega explosion. We managed to snuff out the jet of flame with damp sailcloth before the wooden roof trusses caught light.

I had ridden a pony when I was four years old and was, therefore, relaxed about the prospect of pony trekking with the gang. We rented a dozen horses and off we galloped into the sea at Soar Mill Cove. The horses loved it, but mine didn't love me and tossed me. I swapped horses at lunchtime with the same dire outcome. I walked the horse back to the stables and never went riding again.

I had always had a yen to go to France. My mum was Francophile, though she had never been there. I was drawn to the wild, rocky north Brittany coast, with its 30-foot-high tides. I was equally impressed by the rugged French sailors who came and went in all weathers. Their boats were basic, overcrowded, engine-less and exceptionally seaworthy. After writing several times with the support of Major E.G.M. Pearce, the ICC Secretary, I secured a job with Les Glenans, the world's most prestigious sailing school.

After a long journey by ferry and train, I arrived at daybreak in the bustling fishing port of Concarneau to be met by Jean Bonnet, a boat builder. Bonnet was to become a lifelong friend, but for the moment he didn't speak English, and my schoolboy French was rather light on boat building vocabulary. He offered me a rough Gauloise cigarette and showed me around my new home, a cathedral-high hangar full of enormous sailboats. The top floor housed a canteen and six bare bedrooms. I crashed out behind a curtain in the canteen to wake up to the smell of coffee and toast.

This was not like working in an English boatyard. The men arrived early in their changing room, donned immaculate overalls, and had a glass of rough red wine and a piece of dry bread, and anchovies from a barrel in the corner. At 9am, we all sauntered around the corner to the bistro for a *chouchen chaud* (honey alcohol). Back in the boatyard, I tried to focus to take extra care of my fingers while slicing wood on the whirring bandsaw.

By working my passage as a sailing instructor during the summer, I was able to complete a two-year college course in

Recreation Management. Now 25 years old, this opened the door to a progression of sailing jobs: Derbyshire Education County Sailing Master, Manager Drakes Island Adventure Centre, National Sailing Coach with the Royal Yachting Association, and then CEO of my own company, SportConsult.

As I eased my way towards early retirement, my wife and three grown-up children would not, I guess, have been surprised that I had succumbed to the urge to sail across an ocean, to sail downwind with my own boat, to disappear over the horizon into the last accessible wilderness on Earth. I guess that there is some saltwater in my blood, being a descendant of the John Jameson Whiskey family, owners of three sailing ships in the 1400s, and racing the yachts of kings and the Iverna in the 1890s.

Searching for a dream ship

Step one: I bought Jimmy Cornell's *World Cruising Routes*, the Bible for long-distance sailors.

I realised early on in the search for a boat that I did not have the resources to buy a normal yacht that would be big enough to do the voyage of my dreams. It took a year, and many disappointing inspections of smelly, badly maintained old boats, to realise that the best option would be to buy a second-hand Polynesian catamaran. I was looking for a boat that could sail itself, comfortably downwind, in any conditions of wind and wave, allowing me to sleep, eat, navigate and enjoy the scenery. I imagined spending my days watching the waves endlessly crossing the horizon.

On a cold, wet February day after trudging around Cremyll Creek near Plymouth, I visited Scott Brown, a yacht broker, aboard his barge to see if he had any catamarans on his books that I could afford. As we talked, an e-mail arrived with some great photos of Taraipo, for sale at the right price, in a dream location, a lagoon on an island, in the south of Portugal; a perfect place from which to begin a transatlantic voyage, avoiding the rough English Channel and the Bay of Biscay. Too good to be true. I hired Steve Turner, a yacht surveyor and catamaran builder, and flew to Faro. I was so excited I could not sleep that night. The next day, we took the ferry, in the warm February sun, from Olhao to the island of Culatra and trudged through the marram grass across sand dunes and channels of water emptying on the tide, to discover a drying lagoon full of DIY catamarans. Fritz from Vienna, the owner, had sailed Taraipo from Holland, then decided that she was too minimalist and low on home comforts and was keen to sell.

We climbed aboard over the netting foredeck, washed the sand off our boots, in the bucket provided, and squeezed into the tiny starboard cabin. "Oh dear, this is really too small." I was wedged between two big men, both trying to sell me this boat. Fritz proceeded to list all of the faults with the boat. I got cold feet and decided not to buy. Having paid Steve, the surveyor, plus our air fares and hotel bill, I was pretty down as we awaited the last ferry back to the mainland. Over dinner that evening, Steve, full of good sense, proceeded to analyse my reasoning and concluded that we had been aboard the boat of my dreams and that we should return to the island the next morning. Fritz was pleased to see us, and we proceeded to sign an 'agreement to purchase subject to survey'. It was only at this moment that Steve was allowed to commence his survey. Knowing where to look, he quickly discovered some bits of wood rot, and that the engine was misfiring and difficult to start; nothing that couldn't be fixed. We agreed on a price, £12,000 (the price of a second-hand car), and left for the mainland and a stiff drink. I was far from sure that I had done the right thing.

I still had work contracts to fulfil, the next being in Thailand, where I was to train sailing coaches at the national sailing centre; an intriguing prospect. I had no knowledge of the Thai culture and was not aware that the ageing king had raced sailing dinghies. The plane landed on time at Bangkok, where I was supposed to be met. There was nobody at the gate. As I did not speak the language, couldn't read the alphabet and did not have the contact's name, phone number or address, I felt worried. I sat down and waited. To my relief, a high-ranking naval officer and a matelot walked straight up to me and introduced themselves. Well, I was the only beard in the airport. He explained that I was to stay in a nice hotel in the centre of the city for two nights and that my personal assistant would collect me the following day and show me the city. After a copious breakfast and a read of the paper, I was surprised to be approached by a woman in uniform – my PA and a matelot chauffeur. We had a lovely day doing the sights in VIP style, the car having flags on the bonnet. We had lunch in a floating restaurant – not the matelot; he had to mind the car. The next day, we drove a long way south to the Naval Sailing Centre, where I was accompanied from dawn 'til dusk by my PA/translator. What a privilege to be working with such a disciplined, hard-working and respectful group of sailors. While waiting for the plane home, I noticed a neat electric shaver 'on sale'. I bought it and shaved off my 25-year-old beard. A life changer; I was suddenly ten years younger, though most of my family and friends were not impressed by my fresh, young chin. In fact, most of them didn't see that there was something missing. I had never realised what a turn-off a beard can be to many women.

Back home in Dalgety Bay, I sorted out the essential belongings I would need to take to the boat in Portugal. Two windsurfers and three sails proved a challenge to the airlines and taxis. At Glasgow airport, I tried to pass off the long mast as a fishing rod. The girl at the desk remarked that I must be after some big fish, and with a chuckle checked it through, together with my short board. Friend Duncan and young son Cammy flew out for a trial sail, bringing my long board with them. They didn't have a problem checking in, but we did have a laugh finding a taxi at Faro airport that was prepared to take a 2.75m windsurfer to our hotel, from where we could carry the board to the ferry and then walk it across Islas Culatra to Taraipo, my new ship.

Our first attempt to sail Taraipo was a bit of a disaster. It was breezy, and we were constrained by unfamiliar narrow channels, sandbanks and tidal streams. We nearly blew the sail and mast off the deck, having failed to control the sail as it swung across with a gust of wind from behind. The whole raft-like structure of the boat felt soggy at the joints, and we discovered that the nuts on the 32 bolts that fasten the cross beams to the two hulls had not been tightened. Of course, the faulty injector on the engine and the cooling water pump did not manage to repair themselves while we were sailing, so we wobbled back into the lagoon to lick our wounds.

The tidal lagoon on Culatra is a special place, a haven for many catamarans and impecunious nautical dropouts from society. 'Tidal' means that all of the water drains out of the lagoon at great speed through a narrow, shallow and challenging channel every 12 hours or so, leaving all the boats high and dry.

It was wonderful and a bit eerie to start living alone aboard my acquisition, a strange 'Kon-Tiki' raft-like vessel. The boat needed a lot of repairs and modification to make her seaworthy.

Taraipo, 'a voyaging canoe in a Maori fable' that Polynesians we met said 'was built in the night', is a Wharram-designed 'Tangaroa mk1' catamaran, beautifully home built by Pieter Scargo in Holland. The dimensions are 11m long x 5m wide x 60cm draft. She was made from sheets of marine plywood, epoxy glued to a wooden framework, and skinned externally with fibreglass. There are no metal fastenings apart from the stainless steel brackets attaching the beams to the hulls. Modifications to the original design included a sloop rig, an engine pod, a hinged propeller shaft and aft-facing entrance hatches.

There were no mod cons aboard: no TV, no fridge, no dish-washer, no mains water, no mains gas, no mains electricity, no internet, no mobile phone at sea, no mirror and no shower.

Whenever it was windy and the tide was in, I took time out to scream up and down the lagoon on my short windsurfer, until my arms were pulling out of their sockets. Happy days and lonely nights. Getting the boat ready was a big challenge and I had neither the skills nor the knowledge to resolve many of the problems. Yes, I had done a lot of sailing, but I had never been a 'live-aboard'. The boat needed to be prepared to cope with any situation that could be imagined: capsize, fire, explosion, dismasting, a lee shore, storms at sea and anchored for long periods in remote bays, without power or fresh water. I also needed to have enough to eat and drink for passages of up to six weeks drifting across the ocean. Sufficient electricity was needed to power the autopilot to steer the boat 24/7, the GPS to fix my position and the navigation lights to be seen at night.

Sun awnings needed to be designed and made to protect me from the sun, and rain squalls from behind. I had to paint the topsides and learn how to roll on thick antifouling paint below the water line, between tides and rain showers. Fortunately, Fritz had left aboard a full set of hand tools and power tools, including a 'winkelschleifer' and a wee Honda generator to provide the necessary 240 volts AC. I was intrigued by the 'winkelschleifer' angle grinder, a tool new to me, which came without instructions. I bolted on the metal cutting disc as best I could, but, not being too sure about this, I held it at arm's length over the side and switched it on, and the disc disappeared into the sand at 12,000 revs. Nobody died! I set to with the support and knowledge of Jerry and the other highly experienced sailors who were moored forever in Culatra for one reason or another. A crew to work on the boat and help with the daily chores would have been ideal, but I couldn't find a volunteer to share the very rudimentary existence. I did manage to persuade a few girls I met on the plane, but two nights was enough for them. It was cold at night; two sleeping bags each were essential. The pump-out toilet could not be used in the lagoon, so a trip with a trowel, down a ladder off the bows, to find a bush in the dunes was the order of the day or night. There was no TV for the long nights, little space in the tiny cabins, sitting room only, and no power could be generated from the water-cooled diesel engine when the tide was out. Shopping required a walk across the dunes to catch the ferry to Olhao, a busy fishing port where everything could be found: gas, chandlery, materials and food. There was no fridge aboard, as it would have used up too much electricity. Water was transported in 40-litre plastic containers from the mainland in friend Jerry's fast tender. I soon got used to a bucket wash in sea water. The domestic chores were frustratingly time-consuming.

The neighbours were very generous with their time and their expertise. I learnt how to remove engine injectors, measure electricity, install wiring, laminate with epoxy, and source worldwide weather forecasts and cheap charts. Jerry, my next-boat neighbour and catamaran guru, told me that it was cheaper to buy boat gear over the border in Spain. I ordered a jib furler that I thought essential for single-handed sailing (the jib furls around the forestay rather than having to be lowered and bagged at the end of each day). I sailed down to Isla Christina via Ayamonte, where I was faced with a lee shore manoeuvre to park the boat in a marina berth, which I was pleased to accomplish without assistance (siesta time). Later, the dock master told me I had to move to an inner berth between two rows of immaculately varnished yachts, which I accomplished by 'ferry gliding' in stern first, much to the consternation of several yacht owners standing by on their polished transoms, ready to fend off!

Part Two: Learning the hard way

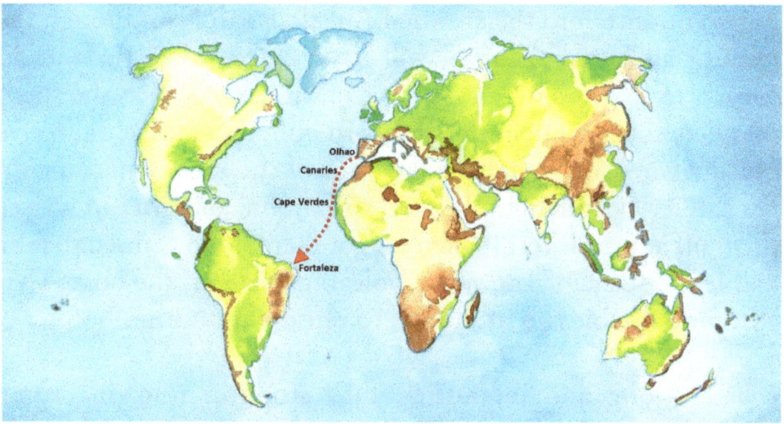

Friend McLaren and his daughter, Megan, volunteered to do the shake-down sail from Portugal to the Canaries. What a nightmare trip that was. It started with an expensive lobster meal the night before we left Tavira, which made McLaren very poorly. It was a rough passage, square tacking to windward. After two nights, the electricity ran out – not surprising, really; we were stupidly using the deck light a gogo and the radar to scare away ships, the off watch were reading books all night, the autopilot was steering, the navigation lights were bright and the radio was blaring. Far too much power was drained from the one old car battery. It was not the downwind passage of our dreams.

We stopped in Porto Santo, Madeira and licked our wounds. McLaren covered the hole in the ripped-by-the-waves trampoline foredeck with some old blue fishing net we found in a rubbish skip. This temporary repair required half a day of surgeon's ingenuity (McLaren) with a sailmaker's needle and thread. Nobody fell through the hole while changing foresails during yet another rough overnight passage to Lanzarote, during which the engine exhaust system, which is lashed under the aft trampoline, was smashed off by a wave. We nearly choked on the acid, black exhaust fumes as we made our dramatic and noisy entrance into the crowded Arecife Porto Nao through the lines of anchored yachts. I will never know what 16-year-old Megan really thought about the trip, much of which she spent huddled in her bunk in the starboard hull, though she did revive quickly when we went on a day tour of the lunar landscape on Lanzarote. Fortunately, I went with John and Megan to see them off at the airport. When we got there, John realised that he had left a bottle of Lanzarote 'volcanic' wine on the boat. That cost me a double run in the taxi to retrieve it for him.

I had a long and interesting DIY stopover resolving some major technical issues, including replacing the dud battery, climbing the mast to change the broken stainless steel forestay tang, fitting a new exhaust/silencer under the deck from the rubber dinghy, re-welding the stainless gooseneck and replacing wooden boom end plugs. The local sailmaker rebuilt the stainless mainsail headboard. Repairing stainless fittings, which can fracture without warning, is a recurring bugbear on a yacht.

I was gifted a neoprene water tank by a neighbour, which fitted neatly into the deep V-shaped bilge, and I in turn donated to the local lifeboat a vacuum cleaner that Fritz had left aboard. I had met the cheery lifeboat coxswain helping a young French couple diagnose the cause of their engine failure, having saved them from shipwreck the night before.

Taraipo was saved from destruction at 10am on a windy morning. I looked out and realised that we were adrift, travelling fast down harbour dragging the anchor through the mud. Kiwi neighbours saw us go by and gave chase in their dinghy. As I drifted close by a French yacht, the skipper caught my line and quickly made fast; the rope stretched noisily, then all came to a halt, phew! I pulled up the anchor warp to discover that there was no anchor on the end; the warp had been severed by some rusty scrap iron on the harbour bed. The Kiwi crew came aboard and we prepared a game plan: Kim stayed on the foredeck with her gloves on to help me lay the kedge anchor while Tony towed us back upwind. An English yacht, Folly, having heard of my plight, offered me an unused anchor that they had spare. We laid two anchors and replaced the warp with an all-chain rode. Another lesson learnt. A Brazilian couple on the last leg of their five-year Atlantic circuit suggested that I would have much more fun if I headed for Salvador da Bahia, their home port, for Carnaval instead of following all the other yachts to the Caribbean. They accompanied me to the photocopy shop to copy their landfall charts of the entrance to Salvador and offered to give me up-to-date charts and Sailing Instructions for all of Brazil on arrival. I had an excellent fixed GPS, but did not have a computer or electronic charts, though every other skipper insisted on giving me a CD rip-off of a world portfolio of 7,000 charts, worth a fortune, which I could not use without a laptop. In order to test all of my modifications and repairs, on a windy day, Tony and I sailed around to the old harbour inside the Islote de Fermina, where I picked up a mooring for the night by diving 2.5m to tie a warp to the ring on the top of an old stone block; unique, but standard practice here – no anchoring allowed. I dropped Tony ashore, checked the weather forecast and settled in for the night. Horror of horrors, at dawn, the wind had changed direction without warning and we were on a lee shore! The old circumnavigator on the boat next door was diving down to untie the mooring while his wife drove gently forwards. I was stuck, as there was a boat close behind and, by the time they had left, the wind was howling. So there I was, alone, like an idiot, pinned close to the harbour wall on a falling tide. I dropped an anchor in case the mooring broke, and watched in awe and disbelief as the waves started to build and roll in from seawards and took occasional calls on the VHF from my Kiwi friends checking that I was okay. Luck was in; late in the day, the wind died and I headed off for Gran Canaria, arriving at Las Palmas in the night. I could not find the yacht anchorage on my radar. Motoring around in the dark, I was suddenly picked out in the spotlight of a coast-guard cutter, like a rabbit in car headlights. They pointed me in the right direction and explained that my photocopied chart was out of date. The reason for coming to Las Palmas was to find myself a crew, as rumour had it that there were dozens of dreamers hanging around hoping to find a berth on one of the 200 yachts in the marina waiting to start the annual transatlantic rally. I have an aversion to marinas; contact with other sailors is not easy there and they are expensive. There, the free anchorage was conveniently just outside the marina and close enough for me to poach a crew. I phoned Montse, an old flame from Barcelona, a cool sailor who was currently chief steward on a ferry in Las Palmas. Sadly, she was not interested in sailing away with me. However, an enterprising Swedish woman had fastened 'crew available' notices to lamp posts around the marina. She came for a trial sail, after which she declared, "I like you, but I don't like your catamaran, which could not compare with my own yacht." She had owned a Dufour 35' keelboat, 'an excellent boat to windward'. There was a bit of an awkward silence as I ferried her ashore and, later, I texted her saying that I had decided to continue the voyage alone. To get ashore to the shops from the anchorage involved inflating and launching my leaky dinghy, carefully attaching the brand-new Yamaha 1.5hp outboard engine bought in Olhao and heading for the sandy beach in the middle of the city. On the way in, I was flagged down by a

Croatian lad on a tiny 26-foot home-built catamaran who was looking for a lift, as he did not own a dinghy and was fed up with arriving wet in the glitzy shopping mall having swum ashore!

When I could see from the esplanade that the waves on the horizon had calmed down, I bravely sailed away on my own, cheered on, and hooted, by Anne and Nev on Peace IV, a luxurious new 42' Polynesian cat that they had spent the last five years building in Bristol, and were heading for the US.

The first gale

The weather forecast was okay, so I was surprised to be caught out by a gale the following day at sea! The wind started blowing me about the deck and the waves started to break. This was the first stability test for the Polynesian catamaran, which the designer says will not flip. In a panic, I took down both the mainsail and the jib, battened down the hatches and lay a-hull, sitting inside the port cabin praying that the boat would not capsize. For 12 hours, I watched the drift track on the screen of the GPS. First north-east (back where we had come from), then east (towards Africa). The wind died and, that night, I found myself in soundings (shallow water) near a fish factory ship in a blaze of light. As dawn rose, I was surrounded by wooden fishing boats, invisible at night on the radar, their only illumination being the occasional click of a cigarette lighter. Then followed a romping sail to the Cape Verde islands; I was surprised to see them through the binoculars high in the distance long before they came up on the radar. Approaching Palmeira on the lee (west) side of the island of Sal, I could see that there was big surf inshore. Weird, as the strong wind was blowing offshore. I was tired and was expecting to find a nice quiet anchorage, sheltered from the steady trade winds, laden with red Saharan sand. I prepared for the worst, closed the hatches, put the spare GPS in my jacket pocket, donned a safety harness and headed for the anchorage, which seemed, from my old chart, to be sheltered by a harbour wall. As I got closer, it appeared that the waves were only breaking along the shoreline, so I took off my sweaty waterproofs and continued my approach, eventually arriving in a very crowded anchorage, no spaces, and the swell breaking on the beach behind. As I chugged into the shallows, a young French couple saw my dilemma and leapt aboard as I passed by to give me a hand with finding a space and carefully laying my anchor; cool. They brought cold beers, and followed up the next day with a Christmas party a trois, sheltering from the howling, sandy wind under Taraipo's awning. I decided that it would be better to sail the Atlantic two up, so I e-mailed Kristina in Sweden to see if she was still up for the trip. She arrived a few days later, at midnight, complete with her belongings all wrapped in cardboard boxes. She was keen to leave at 6am for a shakedown passage to Mindelo. When we made landfall after a night at sea, she was looking pale in the morning sun as we passed between high, rocky islands, out of control, dodging small fishing boats anchored in the stream, the current being faster than our boat speed. It was party time in Mindelo. While ashore buying fresh fruit and vegetables at the market, our inflatable dinghy was sabotaged with a sharp pin. Fortunately, the head of the outboard motor was just above water and we managed to get back to Taraipo pumping all the way.

Dismasted

We set off for Brazil in a howling sand-laden wind; the boat and all the ropes were plastered with red Sahara sand and the visibility was less than half a mile. It was reassuring to have a meticulous and experienced navigator with whom to share the trip, and all went well until we were exiting the doldrums, an area of calms and squalls south of the equator, when we were caught in a sudden squall with the mainsail up. Kristina was steering by hand to keep control as the boat surged forwards at 10+ knots. As I struggled to drop the fully battened main, the mast broke and the whole shebang lay across the deck and on my two stacked windsurfers. All was quiet apart from the grinding of the heavy mast on my fragile windsurfers and the occasional thump of the mast top swinging under the hull. We had dismasted and needed to get rid of the wreckage fast before it holed us. The expensive wire cutters that came with the boat did not work. I tried to hacksaw through the stainless steel shrouds to no avail, finally resorting to slowly unfastening the nuts and bolts securing the bottle-screws. In hindsight, I should have tackled the job with the wonderful 'winkelschleifer', but I did not think of that at the time.

Dismasted

Jury rig

There was a big residual swell running. I was relieved to be on a stable catamaran as we went about rescuing the mainsail from the groove in the mast hanging over the side. It was a great relief when the last wire let go and the heavy rig, together with the radar, deck light, navigation lights and the VHF radio aerial, disappeared 400 metres into the depths, releasing Taraipo to drift freely once more, 400 miles from Brazil. We set about constructing a jury rig to enable us to continue downwind and did not consider triggering the emergency position-indicating radio beacon (EPIRB). Our pride would not allow us to expect or ask for assistance. We were not in a life-threatening situation and there was enough fuel in the tank for the last 100 miles as we neared the coastline. With block and tackle, we were able to straighten the stainless steel tabernacle, set upright the stub of the mast and insert the boom into it. We then had to devise ways of fastening to the masthead two rope shrouds, one forestay to hold up the mast, and jib and main halliard blocks to hoist up the sails, a VHF radio aerial, and a masthead light on top. Having dismasted at 10am, we were underway again by 4pm, flying the tiny storm jib and a horizontal genoa. It's amazing what a bit of adrenalin can do. Kristina had been studying what our options were, as it was not possible to continue, against the wind, to Salvador da Bahia, the entrance to which we had charts for, nor could we turn right for Trinidad, for which we did not have charts. We did not have a chart plotter or electronic charts. I was sure I would find a replacement mast there. Trinidad was too far away; we might have run out of food and water before we got there. We had no detailed charts of the Brazilian coast, so we headed for the only place that appeared to be dead downwind of us, a dot on the large-scale chart of the South Atlantic. We entered its co-ordinates into our trusty Furuno GPS and headed 400 miles at 3.5 knots towards Fortaleza, guessing that there would be a port there and hoping not to surf onto a sandbank before we arrived. We timed our approach to explore the last 30 miles in daylight. Unfortunately, the visibility was poor rain and mist hiding the dozens of windsurfer-lookalike Jangada fishing boats. We followed any of them that were heading towards our mystery waypoint. Amazing. "It looks like the Miami skyline; Fortaleza is enormous; the water is getting shallow; is there really a harbour here?"

"The Jangadas look like surf boats; maybe they beach." Then we saw wind generators, jetties and an anchorage full of Jangadas. Relief all round as we anchored amongst them to cheery waves from returning fishermen as they sailed close by, some shouting, "don't leave your boat, watch out for looters". We chuckled to ourselves, "our raft looks a total wreck, not worth a dime, unlikely to attract attention." The surf on the beach would have sunk our inflatable dinghy, but we had to get ashore somehow to check in with Customs & Immigration. Next morning we headed for the only tower block that had a small yacht, a J 24, moored in front of it. We landed on the steps, to discover a luxurious but deserted yacht club, with a guard at the street door exit. He ordered us to leave our backpacks with him. Confused, we kept our bags, and walked off towards the Customs office passing in front of plush apartment blocks, some with burly bouncers in the doorways. Error! Next thing I know I am was being dragged to the ground by a tough young hooligan trying to steal my backpack. Kristina screamed and laid into him and he ran off. Phew, welcome to the third world. "Kristina, you saved my life." We reached the Coastguard offices, bruised and shaken (not Kristina; she's a very cool woman). What happened is not surprising if strangers are stupid enough to wander along a road that separates the very rich in their condominiums from the extremely poor in a fishing village on the hill. We began to worry about the boat. Having checked in, we did as advised by the Coastguard, took a taxi the 400 metres back to the yacht club and re-anchored the boat close to the security lit police boat jetty. As night fell, I was cooking and Kristina was dozing on deck. I popped my head out of the companionway, a wine bottle in one hand and a glass in the other. Christ, there was a guy with a kitchen knife standing over the sleeping beauty. I lurched towards him, then he leapt into the water, followed by his pal! Argh, now we were scared; we guessed that it was the same youngsters who had failed to get us in the morning. So never mind the supposed safety of police cover, we were going to anchor a mile offshore, 'a long way for the boys to swim with their mum's cutlery'.

Around midnight, we finally had a friendly response to our panic VHF calls to the Coastguard, Police, Customs and Immigration, and the Harbour Master, from the Pilot boat! The pilot explained, "There is a small, guarded yacht marina about two miles further up the coast; it will be easier to find it in daylight." We chugged in on one cylinder; the strong wind was blowing us onto the visitor pontoon. The dock hand helped fend off and make fast, and explained that he would help me anchor and reverse into another berth. I tried to explain that such a delicate manoeuvre was beyond my capabilities in the prevailing conditions. To no avail, he hopped aboard, cast off and told me where to drop the hook. Kristina was standing on the hot, rusty, pontoon ready to catch a warp. I was driving full astern when I saw a cloud of black smoke from the engine hatch. I switched off the engine and we started to drift into the next yacht! With practised aplomb, the dock hand leapt into the water to swim a warp to Kristina. He had seen it all before and dried off in minutes. The reason for the engine overheating was that a plastic bag had been sucked into the water-cooling intake.

As I started removing the jury rig, Kristina set off to find a friendly old Jangada sailor in order to source a local rig or a mast. While waiting in the doorway of his shack, she was attacked by a lad who tore off her gold necklace. We had been warned more than once. The old fisherman could but apologise for his neighbours. The local rigs would have been too difficult to install, and the wooden masts too heavy. He eventually found us a section of aluminium lamp post, which I was able to slot into the stub of the old mast. I fitted halliard sheaves, a topping lift, forestay, and shrouds made with galvanised wire and bulldog grips from a hardware store.

Lamp post mast and cut-down sails

Masthead

As the mast was shorter than the original, I had to reduce the sail area by unstitching the bottom panel in the mainsail and buying an ancient 'IOR' bulletproof staysail from a German yacht to use as a jib. The new mast had no sail track. I made hoops from plastic-covered wire and bulldog grips, to hold the sail to the mast.

After two frenetic weeks, we wanted to check out, but were thwarted by officials with guns, who had us chasing around town in taxis to acquire the necessary number of rubber stamps on our documents. In the end, we gave up and left without permission or stamps in our passports, and headed off towards the Amazon delta, passing through myriad oil rigs on the first night at sea. We called into French Guyana on our way to Tobago and Trinidad to find a taller mast.

Welcome to the Caribbean way of life: lots of bureaucracy, hangin' around and rubber stamps, ad infinitum. It is an overnight sail from Tobago to Trinidad, close-hauled on port tack. We had heard of strong currents setting onto oil rigs. At 2am, I was a bit perturbed to see bright white lights to leeward and, getting closer, through the binoculars it looked like the Eiffel Tower by night. I could not sail any faster and still we were being drawn closer; whichever way I turned, from close-hauled to a dead run, the relative bearing did not change, then I started to make out red and green, port and starboard lights – this apparition was coming towards us! Then I made out another set of navigation lights and started to panic; what was this island of light that was about to engulf us? I grabbed the VHF and called, "Ship, ship, this is yacht Taraipo; do you see me? Over." There was a brief pause, followed by a gentle Scottish voice announcing, "This is lead tug of two; we are travelling at five knots. Do not attempt to pass ahead of us; you could become trapped between the two towlines and finish up under the rig!" We passed behind and she disappeared like a ghost in the night.

In the morning, we made our landfall in the busy anchorage at Chaguaramas, Trinidad. It is deep and the holding is poor, and is surrounded by boatyards and chandleries, with everything a sailor could need, providing his pocket is deep enough. Trinidad is just outside the hurricane zone and home to a 1,000 visiting yachts every summer. You can live cheaply there if you keep an eye on the weather in the anchorage and watch your back when you are ashore. Security is a big issue. The sort of thing that happens is a yacht crew arriving from abroad takes a taxi from the airport. He sits in the passenger seat and, at the first set of traffic lights, the taxi is cut off by another car and stops. The taxi driver deserts, and the attacker takes the wheel, speeds away and crashes off the road at the next corner. He then runs away, leaving our man unhurt, shaken and still strapped into his seat when the police arrive. Welcome to Port of Spain.

It is steamy hot all day long, so it's hard to get anything done, apart from spend money in Budget Marine, which is air conditioned. I was running on a very tight budget. Kristina jumped ship shortly after our arrival. She was not too keen on sharing our basic victualling costs.

It took me a while to visit all the boat yards to secure the best deal before hauling out at Power Boats. When I positioned Taraipo in the slings of the Travelift, the driver refused to lift her and went to fetch the yard manager. The driver feared that, when the slings took the strain, the V-shaped hulls would squeeze inwards and break the laminated wooden cross beams into matchwood. The manager reassured him that, as long as the mast was up and the shrouds were tight, this could not happen. Nevertheless, he told me that, if Taraipo did crumple, her removal would be at my expense. I stood clear of the mast as she lifted, creaking and groaning, and was trundled and chocked on dry land, ready for a repaint, antifouling and a replacement mast, which I was still to find, and never did. The next day, I noticed that the propeller had been stolen off the shaft while I was away shopping. After alerting the security guys on the gate, it turned out that the engineer who was supposed to remove the gearbox for repair had taken the prop to be polished at great cost! I thought my luck was in when offered an old catamaran mast for free; that was sadly too big in section. Eventually, I found an oil rig service company with a heavy gauge aluminium tube just the right internal diameter into which I could slide and pop rivet the lamp post as a top mast. I had lots of help and tons of advice from fellow sailors, and the social scene wasn't bad either; near to Taraipo was the open-air Sails restaurant and bar, where we would have a roti and a beer for lunch, and the weary boat owners congregated for a sundowner and to recover from the heat. At lunchtime, the boatyard guys played cricket (the national sport) between the boats.

Brian, the owner of Magic Carpet, a tiny 26' yellow Polynesian catamaran he had built and then taught himself to sail on his maiden trip from South Africa, had a problem. His boat had been impounded by the boatyard where he had left her three years previously. He could not afford to pay the accumulated fees to launch her. No cash, no splash. It would be a loss maker for the yard if they had to scrap her, so a deal was struck whereby she was launched secretly at dawn. I was in the dock waiting to tow her away with my 1.5hp inflatable tug. Fortunately, there was no wind. The other member of our threesome was Sofia, a yoga teacher from Australia. Her skipper had left her to look after their yacht, and asked me to look after Sofia while he was away delivering a yacht. I was happy to oblige. Firstly, I had to fend off Brian, who thought we could share. No way; she was wee, fun and fit as a fiddle – a health freak. She was even planning to go on a blood-cleansing programme! There were thundering steel band yards to search out at Carnaval time in Port of Spain when the whole town beats and dances for a week. Not for the fainthearted. Keep your hands on your wallet and watch out for Sofia with all these burly Rasta guys soaked in rum on every street corner. Not that I would have been any match for the Rastas. We had a great time together doing all the things couples do, but single sailors don't, like going to the cinema, jazz concerts, lunching out, sailing around the bay, chatting all night and exercising in the cool early mornings. I was quite keen, but Sofia was preoccupied with her yoga groups and her varnishing.

The day I relaunched Taraipo, the yard guys topped up my fuel tank with petrol (gasoline) by mistake. Fortunately, one of them kindly volunteered to come back after hours to help purge the engine and reassemble it, enabling me to escape from the fuel dock.

I was beginning to realise that living the do-it-yourself hobo lifestyle of the tropics was incredibly cheap. I could survive on $8,000 per annum all inclusive. Maybe I didn't need to re-cross the cold Atlantic to Scotland. I couldn't resist the temptation to explore the idyllic coastline of Venezuela; I had to go there.

On my own again – towards Panama

Dealing with Trinidad Customs was a challenge. It took a morning or more to do the documentation, and yachts must leave immediately, which obliged them to sail through the pirate-infested waters of Venezuela in the dark. If a boat anchored in Scotland Bay in order to make an early departure the next morning, the flying squad patrol boat arrested it and towed it back to Chaguaramas, and the owner was fined.

I headed offshore for the night and homed in on the Los Testigos islands before dawn. Dangerously, I confused the only lighthouse on the island with an illegal masthead strobe on a yacht at anchor. Venezuela is a seriously dodgy coastline; the fishermen are wild, tearing around the ocean in their high-powered pangas – one needs to be on one's guard in every anchorage. Being single-handed is that bit more exciting, as you know that you will be quickly overpowered if the *bandidos* ever come aboard. The anchorages are stunning, but, if you are on your own surrounded in a rocky cove, watching the sun go down, you try to prepare your defences just in case you are boarded. I rigged up a few trip wires; loose tacks would not work on the slatted decks of a Polynesian catamaran. I had a red hand flare in my bunk with me so that I could come out blazing if I had boarders. This never happened. I did not carry a gun, as I know I do not have the reflexes of an experienced gangster. As a youngster, I owned a .22 rifle and was one of the few Boy Scouts to have been awarded the Marksman Badge! A boat next to me had its dinghy stolen in the night by desperate and ingenious thieves. We were anchored a long way offshore in the 'Parque Nacionale'; the night was calm and he had a guard dog on deck, and my nerves were jangling. The thief swam quietly out with a light fishing line, attached it to the dinghy, cut the dinghy painter and quietly swam away back to the mangroves to haul in the dinghy, which he abandoned and disappeared with its prized outboard motor.

I was boarded by the 'Guarda Costa' in daylight; they checked my papers and cheekily asked for a beer. When I showed them that there was no fridge aboard, they got the message and left empty-handed.

There were about 70 yachts anchored in Isla Marguerita. If you arrived in the night and left your dinghy in the water, it would be gone in the morning. The 'yachties' organised a VHF radio vigilante service to keep the marauding locals off our boats. Every night, two yachts volunteered to keep watch all night, which I could not do by myself. We all kept our radios on for an attack alert, when we would leap on deck and switch on spotlights and deck lights. "If you are ever attacked, do not resist." One of my neighbours had his fingers chopped with a machete and finished up in hospital. We subsequently made a mass protest at the marina office, attended by the mayor and the police. There followed maritime police patrols for three nights, after which the bandits took over again.

The most sleepless night I ever had was inside an anchorage surrounded by gigantic boulders. I had arrived mid-afternoon and anchored, and was alone in the calm, surrounded by silence. My reflex was to inflate the dinghy, get ashore and make contact with the locals. Once ashore, I wandered nervously through narrow alleyways between the boulders. The first apparition, a blockhouse, was locked and barricaded, and the roof was covered with embedded broken glass. Next, a few shy kids, then a beach bar. It's a Wild West scene. I bought myself a 'Polar' beer and waited for the inevitable inebriate first approach from one of the locals to break the ice. It did not take more than a few minutes, leading to some very friendly chats with the fishermen. I could not understand why the pangas drawn up on the beach in front of the bar were without engines. They explained that they were under constant threat of invasion from marauding bandits, and had to unbolt the heavy engines every night and lock them away in the blockhouse. The next day, as I was about to haul up by hand the 30 metres of chain and 15kg anchor, a fleet of big fishing boats from the neighbouring village came charging in through the narrow entrance, with music blaring, and laden down with families and the priest, who were coming to join a sea harvest festival. They were all the best of friends, it seemed. I decided I had better escape while the sun was still shining.

Later in the port of Sucre, I discovered that an envelope full of dollars, which I had hidden in the battery box, had acquired a 'waterline mark'. On closer inspection, it was not a waterline; it was an acid line. There must have been some spillage from the batteries when the boat was going fast in rough weather. All of the greenbacks had become brittle, so I sought to offload them as fast as possible at a currency shop. The big cashier lady let me in, locked the door behind me, sat behind the counter, took the envelope and roughly withdrew the dollar notes, which disintegrated before her eyes. She shrieked, grabbed a gun from the desk drawer and bundled me out of the door with my monopoly money. I got the bus back to the marina and set about reassembling my greenbacks with Sellotape. Some weeks later, an American sailor friend offered to take them back to the States to get them refunded by the Bank of America. She kindly deposited the cash in my bank account about a year later; lucky me. The other problem I needed to resolve was an autopilot self-steering arm malfunction. An Italian on the boat next door showed me how to dismantle it. The little drive motor was burnt out. "A cheap replacement can be found in any country in the world; they are standard parts in most fruit machines." Job done, with some careful soldering. I now carry a spare autopilot for break-down-at-sea situations.

From a sailing point of view, the coast of Venezuela is nigh on perfection. The wind and current push you firmly downwind and there is no way back. There are pristine anchorages, but approaching gaps through reefs, downwind, requires pinpoint positioning or you are wrecked. In Puerto la Cruz, I had the galling experience of seeing my good friend, Christian, who had helped me to paint Taraipo in Trinidad, selling off the bits and pieces that he had managed to salvage from the yacht he had built and sailed for many years. He had fallen asleep on watch while hove-to, to windward of the airstrip on the Isla la Tortuga while waiting for dawn to break, having got there too early on the wind and current roller coaster. Three weeks after he crashed ashore, I was able, sadly, to walk through the side of the hull. The islanders had kindly helped Christian remove and store the masts and all of his sea-soaked possessions.

The sad thing about Puerto la Cruz, a five-star tropical Venice, is that, when you go past the guard and out of the gate, you have to get into a taxi or have a good chance of being mugged. I could not resist the thrill of chancing my arm; walking to the shops, I would dodge into the next bar if I sensed that I was being followed. Every customer at the yacht chandlery shop was locked in and locked out. They would not allow me to walk back to the boat and gently insisted on taking me by taxi. Now that is what I call customer care. It was not surprising when every man and his dog knew that we were exchanging foreign currency there for US dollars at ten times the rates that the high street banks were offering.

Los Aves, the last stop in Venezuela, was a dream unrealised. I imagined that I would be able to sail fast on flat water for 10 miles or more on my windsurfer and get back. Though the massive lagoon looks awesome, when you are on your own, it is difficult to find an anchorage that you can get back to after stretching your biceps windsurfing for a few hours.

I had planned to leave Los Roques at daybreak to avoid an overnight sail to Los Aves, the islands of the blue-footed boobies, but had to up anchor in a panic at dusk because of the vicious mosquitoes in the final anchorage in Cayo de Agua. There was lots of lightning at night over the mainland; a threat to yachts with metal masts. If struck, the power is paralysing, as it burns through all the wiring and instruments, causing fire and shock.

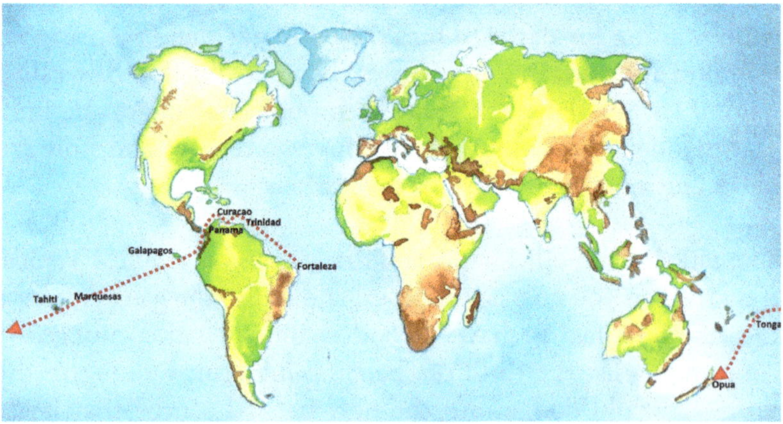

After several weeks of solitude and despite a bit of a bump with another yacht while anchoring under sail in the crowded anchorage at Spaanse Water on Curacao, getting back to civilisation was always good. I could paddle ashore – the boat was safe – have a cold beer in Sarafundy's, buy some water, meet da beautiful people, stretch my legs and sus out the bus system to the city, and do some serious shopping. The next day, I was on the 8am yachtie's radio net hoping to find someone to help me revive the recalcitrant Renault-Couach two-cylinder diesel engine and was lucky to have a Dutch chief engineer come and tell me how to remove a jammed injector, where to take it for testing, and loads of other bits to finally get it going again for the next leg to the San Blas islands. He was a smart guy, a test engineer and troubleshooter on ships all over the world. "John, do what I tell you and call me if you get stuck." After much blood and tears, and oil under the fingernails, I managed to get the thing going again, feebly, with just enough power to chug into and out of anchorages.

According to the Pilot Charts, the passage from the ABCs (Aruba, Bonaire and Curaçao) is always rough. I did not believe this information and hoped that conditions would be lighter for me. I had a few violent nights and, on day two, saw a cargo ship in the distance. Feeling a bit lonely and intrigued to know whether or not my two radar reflectors (one at the masthead and one on the port tiller) gave a good echo… "Ship, ship, this is yacht Taraipo, do you read me? Over." Silence, then a crackled response: "This is British-registered ship BP Caraid, receiving you loud and clear. Over." Bingo. I was pleased that my new radio aerial was working well. "Can you confirm at what distance you picked me up on radar? Over."

"Eleven miles, but we do not have you on visual; are you a catamaran? Over?"

"Why do you ask? Over?"

"We seem to have a double echo." Cool; that meant that both of the reflectors worked.

Arriving downwind in the unlit San Blas islands could only be done in daylight, as the positions from my GPS were about a quarter of a mile out from the charts, so it would have been easy to take the wrong gap in the outer chain of islands and crash into a shallow reef. I got there just in time to find my way into the anchorage nearest to the entrance before the sudden nightfall at 6pm. The next day, I was amazed to see that there were several thatched huts and a heavy, dugout log canoe on the beach about to be launched. The Indians paddled by with hardly a wave as they prepared to step the short mast and sail off to the mainland, with coconuts to trade.

Commuters in San Blas

Every coconut has an owner

When the sun was high enough to pick out the reefs, I eyeballed my way into the next anchorage; always a worrying experience first time around. Somehow, I had managed to find myself in a perfect anchorage for windsurfing in the Holandes Cays, and no other yachts wanting to join me, all preferring to be together behind the reef on the windward side of the island. This left lots of space to rig and launch the windsurfer off the back of the boat, in light airs, and then to sail over the reef and around the island for exciting wind and waves. Windsurfing over shallow coral reefs is very tricky, as you can easily smash off the skeg, rendering the windsurfer virtually un-steerable and compromising your return to base. The other danger is falling off onto the coral, which is lethally sharp and full off poisonous stuff that can quickly infect a foot, hand, leg or more. This is not what you want to happen when you are on your own in a remote anchorage. There are no roads into Kuna Yala, the San Blas islands. There are schools on the bigger islands, but no hospitals. The Kuna Indians are used to looking after themselves and living off the bountiful ocean, and plots of mainland to which they sail to farm daily, in their very unstable dugout canoes. Sadly, they now make lots of dollars from tourists who arrive in cruise ships and yachts. One of the island villages transforms itself into a tourist trap at 9am just before hundreds of cruise ship tourists are ferried ashore and transforms back to normal at lunchtime when the last tourist have left. Yachts can stay for long periods as long as they have checked in with the navy at Porvenir and pay their dues that are collected by each island chief. It's all done properly, receipts are issued, and the dollars collected buy schoolbooks and materials for village projects. The men fish, farm, guard and collect coconuts for export. The women raise the family and do some amazing embroidered 'molas' that are sold to tourists. The Kunas sail out to the yachts daily to sell fresh fruit, veg and bread. One can hang out in the San Blas for a long time, which I did. At Christmas time, I distanced myself from a group of American yachts who organised a kitty in order to have York ham and turkey flown in from New York. I finished up in Kanildup anchorage with one other boat, a young German couple, Johannes and Nellie, independent and fun, and happy to keep an eye out for me when I went off on my daily windsurfing adventure. Taraipo was especially well adapted for the routine of putting together the windsurfer sail, mast and boom on the wide deck, and launching and retrieving the rig and board. A task that is not easy to do on a normal, high-sided yacht.

I got a real buzz windsurfing fast, for endless miles, over sand and coral, the warm water spraying my bare feet. It is addictive to the point of telling myself 'stop after two to three hours' while I still had enough energy to self-rescue should I have an equipment failure, or a catapult fall, wrecking the sail or injuring myself.

Windsurfing bliss in San Blas

On New Year's Day, two outboard motor-powered canoes, loaded with men and nets, beached on the island. The men lay around waiting, watching us and cooking lunch on a fire of coconut shells. Suddenly, there was all hell let loose, the guys falling into their canoes as they launched and headed out towards us, and proceeding to circle laying a long net. There was a lot of splashing with paddles and men in the water to panic the shoal of tuna, which they had spotted from the shore and was now encircled by the closing net. What a good start to the New Year; the boys were elated. It is not every day that a shoal will surface. Later, a scuba diver from the next yacht swam by with a half-dead shark on his line. It took them ages to gaff it, stab it and finally haul it aboard, slit open the stomach and remove his swallowed net bag full of fish!

I felt the urge to move on and set off on an overnight passage 'around the corner' towards the Panama Canal, where I planned to arrive at the entrance, downwind again, in daylight. The only chart that I had was a tracing on greaseproof paper taken from a neighbour's chart. The scenario the next morning was 'dantesque'; it was blowing old boots, white caps on the waves and Taraipo travelling fast through an anchorage full of manoeuvring ships and tugs outside the harbour wall that is miles long, with two narrow entrances. When I thought that I could see an entrance, I prepared for the worst-case scenario of a wave breaking from behind. I closed the hatches, losing access to the GPS and 'chart', and clipped my harness to the jackstay. The visibility was not good in the driving spume, and there were no ships either entering or leaving, which was a bit unnerving. Was I heading for the correct entrance or oblivion? I radioed 'Harbour Control', who advised me to stay as close as possible to one side of the entrance. There was no turning back as I cruised through the gap. Still no ships. The bows buried into the back of a wave and then there was a rattle of anchor chain running over the side. I ran forward to stop it and brought up short at the end of my lifeline. All of the chain ran out, hanging out in a fifty foot loop under the fore-beam, the anchor being lashed on deck. Then all was calm as we drifted slowly and gracefully into an enormous, almost empty, sheltered harbour. I recovered the chain and flaked it ready for anchoring between the other waiting yachts on the mud flats, by the Colon entrance to the first lock of the famous canal. It cost $500 to transit the canal, which I did not have at the time. The next morning, I inflated the dinghy, attached the little outboard motor and chugged half a mile into the Panama Canal Yacht Club, a splendid old, crumbling establishment, to get onto the internet for the first time in months. Lo and behold, there was an e-mail from a cousin in Kenya announcing that my dear godmother, Marian, had left £500 in her will, enough to get me through the canal. If you are an impecunious single-handed sailor, it takes a while to prepare for the transit. You need lots of big fenders for lying alongside tugs or greasy black lock walls, and you have to take on board an official canal pilot and four line handlers for two days. Four extra-long lines, home-made tyre fenders and an engine capable of maintaining a minimum speed of five knots. The pilot has to be given a good meal and access to a clean toilet; no 'bucket and chuck it' for these guys. There is a pre-transit inspection. A buffer fee of $1,000 is swiped from your credit card in case of a screw-up or breakdown in the locks halting the flow of 42 ships per day that pass through the canal. You can either pay local line handlers or share crews from other yachts. I did four 50-mile trips (returning by coach) on other yachts in order to get four volunteers to transit with me.

Friend McLaren joined ship in Panama City. He managed to find his way from the airport, loaded down with kilos of home-made cake. In the heat of the night, he yelled across the water from the top of the harbour wall to wake me up for a lift out to Taraipo. The next night in the Isla de Taboga, we were woken by the anchor alarm. The fisherman's mooring that we thought we had picked up turned out to be a keep-pot, which was dropped quickly as we drifted out into the eerie lights of the big ships at anchor. The next anchorage was not much better; the only other yacht there informed us as we drifted by with the current carefully laying our chain in a depth of 10 metres 'that we should be aware that he had laid 250 feet of chain', meaning that we should go somewhere else. We didn't, and we did not collide when the current reversed in the night. After a tortuous passage through a maze of channels in the mangroves of Panama's west coast to Puerto Pedregal, Chiriqui, McLaren had to fly home.

Into the Pacific - free again

During all of the time in Panama, I had not managed to find a crew with whom to cross the Pacific. I noticed on 7knots.com that my old crew member, Kristina, was languishing in the yacht club in Colon, 450km away. I knew that she would not come with me if I invited her by e-mail, so I decided that ambush was the only possibility. I took the overnight bus from David City, arriving at the yacht club at 7.30am, sat down for breakfast and who should walk in but the lady herself, having already been out for her training run. She was very surprised and took the bait. We cleared her dusty downtown room and caught the midday bus back to Porto Pedregal. She was not enamoured to be back with me, but was pleased to be able to continue with her circumnavigation. She never liked Taraipo either, and was keen to escape to the ocean from the mosquito-infested mangrove channels. I misjudged a reef and we grounded near the entrance. I hopped over the side to push us clear, but the ebbing tide dragged my feet out from under me. I managed to hang on and pull myself back aboard, wait an hour until low water, less current, and gave the antifouling paint below the waterline a gentle wipe.

It was great to head offshore accompanied by an excellent sailor with whom to share the voyage. The bad news was that the wind was light and against us, and the current was dragging us sideways; we almost missed the Galapagos, which would have been embarrassing. We would have run out of food and water before getting to the other side of the Pacific. Unwittingly, we chose the worst anchorage, Puerto Ayora, a lee shore would you believe. It was bustling with dozens of charter boats loading passengers flown in daily from Ecuador. We needed enough water for five weeks at three litres each per day – 270 litres. The only water available was made by reverse osmosis and was expensive. We tried without success to fill our jerrycans in the public conveniences. The other crazy expense was to have the boat 'fumigated'. This required the local chandler to be ferried out with an aerosol full of anti-rat powder, which he carefully squirted into a few corners of the bilge. The next day, I had to collect and pay for a beautiful fumigation certificate and take it to the naval headquarters. We discovered that yachts are not allowed explore the islands unless accompanied by a local guide; not a service that we were prepared to pay for, so we headed off into the wide blue yonder. We were frustrated with each other, no longer the best of friends, and we had a long way to go.

Wow, from dawn onwards, the island of Hiva Oa grew on the horizon, the most spectacular of landfalls, with Atuona the only harbour hidden in a fold in the cliffs.

French Customs and Immigration were always easy-going. We checked in with the gendarme at Hiva Oa and were then free to roam the myriad islands and atolls between there and Papeete more than a thousand miles away. Another plus is that health treatment was free. Kristina, who had fallen down an open hatch, was treated in the Taioha'e hospital on Nuku Hiva for damaged ribs. After the consultation, she was invited to the surgeon's garden to help herself to as much fruit as I could carry! In the Tuamotus, I had to go to a nurse's surgery to obtain free antibiotics to combat a coral-infected wound, which was opening up and getting deeper and deeper into my big toe. Our other 'disaster at dawn' in the approach to Fakarava atoll was to break the mast at the joint that I had made in Trinidad. The entrance under power, with the mast on deck, was tricky and, in hindsight, funny. What happens in the atoll passes is that the tide roars in and out at eight to ten knots, creating giant waves, so you have to time your arrival at slack water when the tide is reversing. We stood off for a while and took a guess as to low tide, only to find ourselves careering sideways through the narrow channel, totally out of control. We did a '360' in the right-angle turn in the coral channel to be spat out into the lagoon. We then eyeballed our way through the coral heads to finally anchor in the lee of a Motu and set to work dismantling the broken mast. After two days, I realised that we did not have enough pop rivets to sleeve the join in the shortened mast, so we reluctantly left our sheltered anchorage to try our luck off the village. The locals were generous; the baker gave us a tank full of diesel. Alphonse, a pearl farmer, drove us 8km around the atoll rim to his workshop to straighten and weld the tabernacle. We stayed for weeks at no charge, in the small boat harbour at the airstrip, where villagers would leave small piles of pop rivets on the wall for us. When the plane landed each day, the pearl farmers powered into our harbour to collect goods and school-children. Most of their materials came on the fortnightly ferry from Papeete. Kristina started her daybreak running routine again. It was too hot to run in the middle of the day and she did some Polynesian dancing, hip wiggling, with the women from the nearby shop. They would glide by in their chrome-wheeled, air-conditioned, automatic four-by-four, steering wheel in one hand and a Coke in the other, smiling big Tahitian smiles, *Tiare* flowers in their hair, while we were struggling in the heat with no awning to shade us from the sun, while re-rigging the mast tube. I had to remove the bent topmast (lamp post), saw off the bent portion and re-insert it into the top of the mainmast. I cut the bent portion lengthwise and used it as a sleeve inside the join. When we came to take our departure, the farewell from the ladies was very emotional. I guess that we were supposed to believe that one day we would return; not something I envisaged in the foreseeable future.

With a shorter, stronger mast and a permanent reef in the mainsail, we headed for Tahiti, the island of dreams, toweringly spectacular, with a shallow and sheltered lagoon. No sooner had we had dropped the hook than Jack and Chrissie from Naga, last seen in Venezuela, came alongside in their dinghy and dumped their old wind generator on the deck and told us we needed one. "True, but I don't have any spare cash." They accepted a plastic sextant in exchange, a totally superfluous instrument now that we had a GPS. We revelled in the variety of products at the Carrefour supermarket while storing up for the passage to the next big shops in New Zealand, thousands of miles away.

The long haul to New Zealand

We survived a heavy gale on the leg from Nuie, to the north end of Tonga, using our technique of dropping the sails and lying a-hull, staying inside, scared and keeping the hatches closed to keep out the occasional rogue waves as they broke over the boat. On the short-wave Pacific radio net, we heard that a fifty foot mono-hull yacht near to us had decided to run south, wiped out and lost his rig, and the island of Nuie was devastated by the same storm. That confirmed to me the seaworthiness of our Polynesian-type catamaran.

Tonga is a sailing paradise, with flat water, plenty of wind and lots of protected anchorages. The locals are welcoming and fun. We broke a stainless shackle while trying to weigh anchor in the coral-strewn main anchorage in Vava'u. Our technique, when stuck, was to haul the chain tight, then run aft, and jump up and down to lever the anchor off the bottom. Fortunately it was shallow enough for Kristina to dive in and tie a rope to retrieve our precious anchor. For several days, we were embayed in a tiny reef-protected pool at Ha-Ano Island well off the beaten track. It was so windy we could not get ashore on the first day. All was quiet in the village; we were met by a few children who guided us to the teacher's house, where we were given breakfast, and I was invited to 'Kava' with the men; a dubious pleasure. What you do is sit cross-legged and ceremoniously pass around a bowl of cloudy water from which you take a gulp. There is a guitar, singing and serious discussion about village affairs. After several rounds, your lips start to go numb. "Time to bail out!" They received Radio Tonga, but no TV, and there was only one telephone hanging on a telegraph pole tended by the kids who sprint to find the person being called. There were pigs and pig shit all over the place. There was a perimeter fence to keep them in, to prevent them getting out and into the surrounding plantations.

In the Ha'apai Group, with Kristina up the rigging, we eyeballed our way through the coral heads and anchored at dusk off a picture-postcard islet with a single palm tree. As night fell, we could see whales blowing and then smell them. As the wind turned, we found ourselves snookered in the coral. We had to up anchor, leave in a panic and guess our way out to deep water. There was nowhere to escape to in the pitch-black night, so I set up a two-mile clear course on our paper chart, between two GPS waypoints. We took turns on watch as Taraipo reached steadily back and forth all 12 hours of the night.

In Nuku Alofa, Kristina jumped ship again. She was determined to finish her circumnavigation within 12 months. I was annoyed, but not surprised; she was a good shipmate and we had had some great times together. She had quietly arranged to be picked up with all her kit by the skipper of an American yacht, leaving me in the lurch. There were no spare crews to be found in Nuku Alofa.

With the arrival of the cyclone season, I had to head south to New Zealand, over a thousand rough miles without a crew. On the radio net, the first yacht to leave warned, "Check the zips on your waterproofs, guys; ours are 'frozen' with mould after two years in the locker and its getting cold down here."

I waited and waited for a weather window of at least a week free of gales before setting off. As I crossed, the days got longer and the nights got shorter, and the forecasts were threatening mayhem. I did not want to be blown away from New Zealand and have to beat back. That would have been an impossible task. So I sailed as close to the wind as possible on the starboard tack until the inevitable depression from Australia arrived. The wind screamed for a day and a half as I drifted away from New Zealand. I was hit by some big breakers, but nothing broke. I kept a lookout from the restricted view inside the cabin, and hooked on and checked on deck after the big hits and any time I felt brave enough to venture on deck.

The last two days of gentle spinnaker run were, in total contrast, blue skies and birds as North Island lifted slowly above the horizon. After one more night-long vigil, the 10-mile-wide entrance to The Bay of Islands opened up and I drifted slowly into Opua marina. The Kiwis were very strict about Customs and Immigration; you had to give warning that you are coming and radio at least 12 hours before arrival. I was handicapped by a defective VHF radio and no engine power. My feeble calls were eventually picked up by friends moored in Russel who relayed my call to Customs. Late in the day, I finally drifted with the current into the Customs quarantine dock. I hopped onto the dock and proceeded to walk Taraipo like a dog on a lead the last 150 metres, at which stage the officers kindly intervened, made fast and helped a very tired man back aboard. It was pouring with rain as we all squeezed into the tiny cabin to complete formalities, which are draconian, and, because my arrival appeared suspicious, they brought aboard two officers wearing overalls and surgical gloves to inspect every corner of the bilges. I was intimidated, four of them and one of me, in a cabin space two metres by one metre, and I was worried that they had been tipped off that I was carrying drugs or something more sinister; who knows? I realised that I was talking non-stop and apologised. "Never mind, sir; most solo sailors are the same when they arrive here. You are clear to go ashore and buy some food for the weekend!" They had confiscated all of the food, including an unopened jar of New Zealand honey bought in Tonga.

Part Three: New crew

My daughter Annick and I had a brilliant time cruising the east coast of New Zealand. The day she left to fly back to Scotland, she said, "Dad, we are fed up with you sailing by yourself; get yourself a crew." Not an easy task in the antipodes. I had searched previously on the internet without success. There were lots of dreamers out there, but, when it came to the crunch, they were not ready to take the leap into the unknown, to leave their comfort zone. If they made it to the boat, after checking out Taraipo and interviewing the skipper, they invariably hurried back ashore, saying that they would let me know; but didn't.

It wasn't that way with Nicole. She was one of a list of women looking for crew positions on crewnz.com. I thought 'that sounds a nice name' and replied. There was no response. I posted 'looking for crew' notices at the yacht club and marina in Opua, and continued preparing the boat for a solo departure. The waterfront was crackling with excitement as several hundred yachts waited for the cyclone season to end and for the first weather window to open, allowing us all to sail north to the hallowed tropics. Weeks later, I received a weak call on my mobile from the Nicole applicant in Australia, saying that she was still interested. I was excited; she had a lovely voice and sounded genuinely keen. Two days later, she was on the doorstep, well at the sailing club steps, with two small bags. Cool as! Neat, bouncy and smiley, and delivered by her six-foot-tall previous skipper. I shook hands and thanked him, and he disappeared from the scene. We carefully loaded the little inflatable dinghy and chugged out to Taraipo, lying to the tide by the far shore, on a calm Sunday evening. I was thinking, *if this one stays, it will be more than a miracle.* She was chatty and fun. We clambered aboard and spread out on Taraipo's wide deck. Of course. everything was neat, tidy and clean. I explained that she would have her own space, the minimalist starboard cabin, which contained a single bunk (with a courtesy curtain I had fitted for the occasion), the cooker and sink, plus food and water storage. I showed her my cabin that contained a single bunk, the chart table and instruments, windsurfing sails, boom and mast, and the toilet (also with a new courtesy curtain).

Captain's cabin port side

Crew quarters starboard side

I carefully explained how to operate the pump in, pump out mechanism, and that 'nothing that had not been eaten' should go into it. All very romantic, but important if you don't want to be up to your elbows unblocking the toilet on a rough night at sea. So we chatted; she didn't seem to be in a hurry to leave. I cooked some spuds, cabbage and an onion omelette (the extent of my repertoire). She was up for staying the night... yes! I poured myself a glass of wine. Oops, she didn't drink. I brewed an Espresso and washed up in her cabin, then left her to turn in. I couldn't believe my luck, sitting under the stars thinking if only she could stay. The next day after breakfast, she still did not want to go ashore. This was very unusual behaviour. New crews normally have the urge to go ashore after a night aboard. In fact, it appeared that she had, with determination, moved in. With no hesitation, I gave her a paintbrush and a tin of white paint and she started her familiarisation with life afloat! I started singing again, something I had not done for a long time. Friends Jack and Chrissie came over to check out the new crew and invite us to dinner aboard his racy trimaran, Naga. Jack was less than impressed when Nicole nearly fell in the water while climbing up over Naga's rounded topsides, though he wasn't slow to grab her by the bum to save the moment.

I had enough food on board for a month. The passage to the nearest tropical islands, Tonga or Fiji, would take about 10 days. We needed to double the amount of food aboard. This entailed hitchhiking 70km to Whangerai. When hitchhiking alone, I would wait about an hour before getting a lift. With Nicole in a yellow T-shirt and me hiding in the bushes, it was 'first come, first served'.

It was a bit weird, Nicole a beginner, deciding to sail a thousand miles across an ocean with an older man. In my wildest dreams, I did not think that there was any chance of a romance. However, we quickly gelled into a seagoing partnership.

Nicole – Swiss 33 years

Coincidences, destiny or maybe because of penguins? Call it what you want. Back in Switzerland, I thought if I didn't go travelling now I would never do it. I quit my job, sold all I could from my flat, stored the small remains my few belongings at a friend's place and went to discover my dream places – Hawaii, New Zealand, Fiji and Thailand. My relationship had just come to an end too. I left Switzerland without any ties and planned to travel for at least a year – if I was careful with my money, maybe two years. After three months in Hawaii, my English was far too good for the course that I had booked in New Zealand, so I looked around for something else I could do and saw in a yacht club a 'learn to sail' notice and signed up for the course. I was in a very nice group – six women and three instructors. It turned out to be more of a survival course than a sailing course – the water was still cold and, one evening, I counted more than 20 capsizes. At least after the first one I knew that I had to hold onto the boat once it was upright before it sailed away; not obvious to a complete beginner.

There was one moment when our little dinghy sailed properly and I thought, That's the feeling I want to have living my life. So I set about finding a boat that would take me to Fiji instead of flying there. I asked around what I would have to look for and a friend told me not to worry about the boat – but to sail with somebody who would be nice enough to be with for a few weeks.

By chance, it was the right time to be in New Zealand – the time the boats head up to the Pacific. "Swiss, 33, basic sailing and advanced cooking experience, wanting to sail to any of the Pacific Islands. Please contact Nicole." By then, I knew the difference between a mainsail (the sail) and a mainsheet (the rope that controls the mainsail), but did not dare to do anything on my own. When I saw Taraipo, I loved it. It seemed to be like a float on the lake of Zurich with two hulls; a big swimming platform. John looked like a nice enough man and he had sailed with his daughter – so I thought he must be an okay man. At that time, all I wanted to do was to go sailing – no relationship, no man. Being 30 years younger than John is, I was sure he was too old for being attracted to. But a boat is quite a small space and I fell in love with John.

In at the deep end

It takes about 10 days to sail to Fiji from Opua. After a few nights, the girl started to feel fatigued. She was finding it hard to be up at all hours, doing two-hour stints on watch. I was in my element, alone on deck, in the middle of the ocean in the night with a sky full of stars, watching out for squalls and ships. We changed the night routine, back to my being a single-handed sailor from 9pm until 5am, with Nicole giving me a spell from midnight till 2am.

Our first extreme weather together was a game changer. When the wind gets up to gale force, survival becomes a preoccupation; you are no longer in control of your destiny. When the wind is screaming through the rigging and rogue waves are trying to smash in the windows, you can't stand on deck. We were scared, locked down in the tiny cabin, with only food and coffee for comfort. I studiously avoided any touching of arms and legs; quite difficult when the urge is there, in such a confined space, with no headroom, the heat, the humidity and the minimalist dress code, shorts and T-shirt. In the night when it was cooler, we only had our bodies for warmth, and the unbelievable, inevitable happened. Her feet were cold, so she slid them under my thigh. I was careful not to notice this at first, and sauntered back to my cabin the other side of the boat.

We shared the small wooden bench, looking out vaguely for ships. The wind and the waves died down quickly and we continued towards Suva, from where she would be flying home to Switzerland. Twenty-four hours before arrival, the electrical self-steering broke down. This was the moment that I discovered that the new crew was a natural sailor, steering a steady course with ease. We took turns each hand-steering until our landfall in Suva, the capital of Fiji. The heat hits you when you come ashore. We quickly went to the fusty old Royal Suva Yacht Club. With the wobbling roof fans whirring above, we sank into the soft armchairs, downed a cold beer and a coffee, and got our bearings. "If it's local information you are after, ask the barman, not the commodore." The bus stop into town was along the road facing the prison gate, its perimeter wall daubed with graffiti. There were no white faces on the bus or in the bustling fly ridden vegetable market; a bit of a shock to Nicole to be in the white minority.

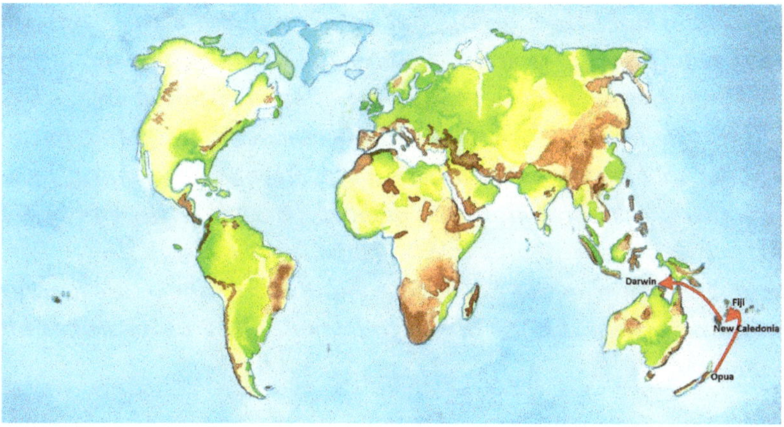

Nicole – a dream come true

I had realised my big dream – I was sailing to Fiji. "If that is possible, everything in life is possible."

The best place to be was on board Taraipo, so I spent the last days in New Zealand painting, drinking coffee and simply enjoying being on the boat. John meanwhile went off to do his tasks ashore.

Before leaving NZ, our joint shopping trip turned out to be a little nightmare – him being indecisive, trying to be too nice. At the end of the day, I was sitting on a bench, waiting for John and thinking that fortunately we only had to sail together; he was not my boyfriend.

John prepared the navigation to Fiji – he drew a line on the chart, straight with a ruler. When we got to Fiji, I thought that's how it was – you drew a line and you followed it. Even when it blew a gale, we still drifted up our pencil line.

My first trip, looking back now, was very pleasant. At the time, I thought it was terrible. We had a gale and laid a-hull. "Is that what you like?" I asked John as we waited 14 hours for the gale to stop. The waves came from behind like mountains and washed between our hulls. Imagine someone slamming a door next to your ear for hours on end and you have a feeling of what it can be like sitting out a gale. Plus, all the tension, the power of the wind in the rigging, and the waves. I felt tense, vulnerable and frightened. I was completely exhausted from the night shifts and, at the end of the gale, I just sat there and cried. "Have a good night's sleep," John said, and that was all I needed. We found our rhythm – me needing much more sleep than he did; both loving coffee and the simple life on board.

After a few days checking in with the authorities, and stocking up with food, water and diesel, it dawned on me that she was not organising her return flight ticket to Switzerland. In fact, you could say that we were having a buzzing time, cold beers in the yacht club, internet access, and all the fresh fruit and veg we could want, and fun locals to chat with. She wasn't going to catch a plane, so we headed off to what I thought might be a romantic, secluded and secure anchorage, and Bequa Island was all of that and more.

Nicole – provisioning

We did not have a fridge. Something I could not have imagined at the beginning. I wonder what makes mayonnaise last outside a fridge for two weeks. Fresh meat, butter, cream and cold drinks was what we could not have aboard. We found margarine that didn't melt away, milk powder, cream powder and good-quality tins of ham and meat. All the fruit and vegetables were not a problem anyway – we bought them in the markets; they had never been chilled, so they lasted for a very long time. Cabbage, onions, potatoes, tomatoes, cucumbers, lemons and oranges lasted for months; carrots and tropical fruit had to be eaten fast.

Market in Tonga

Bananas by the bunch – hard to resist

After yet another scary downwind entrance through an unmarked gap in the coral reef, we were all alone. The initial eeriness soon passed and we were able to relax for a few days and enjoy each other's company in a wild and wonderful place. One day, some islanders arrived with a mile-long net, which they anchored to the shore at one end. The other end was towed in a semicircle around the shoal of fish and back to the beach. They waved from a distance, but did not disturb our idyll. Reluctantly, we moved on to Yanuca island village, hoping to buy some fresh food. It was a nightmare anchorage, full of coral heads, which grabbed our anchor chain so tightly that we had great difficulty unwinding it to get away. Naively, we had brought with us some worn T-shirts to trade or give to the islanders. We were embarrassed to be welcomed by fit young men wearing brand-new Adidas T-shirts.

We chose to head for Vunanui Bay, a mainland anchorage, to buy fresh stores. "You can't live on love alone." It seemed idyllic.

We anchored in the shallow deserted lagoon, a long way from the village, and there was enough breeze for me to go windsurfing. As I returned to our ship, I was panicked to hear Nicole being harangued by an irate villager standing in his rowing boat shouting, "You must give me kava for our chief." Fortunately, we were prepared; it is a tradition to take a gift to the chief on arrival in a new village. Apparently, we were anchored in the middle of their prime fishing ground, and playing around on a windsurfer was not a good idea. Nicole reluctantly gave him the expensive kava that we had bought in Suva. Ominously, he convened us to a meeting with his chief. The next morning, we were welcomed to the chief's house, where we sat cross-legged and were served with tea. Much to our relief, the chief and the men of the village had happily drunk the kava the previous evening. We visited the school and sang with the class, and we invited the chief for lunch on Taraipo. For some reason, I finished up chugging back and forth the one-mile trip to Taraipo in our tiny inflatable dinghy, ferrying the chief and half the women in the village. The chief proceeded to stretch out on deck and start eating, chatting and giggling all afternoon.

In the small hours, a few of the women rowed by and were calling 'Nicole; John' and giggling; they guessed that were not in our separate bunks! They had a baby with them, and they were out fishing all night while their men slept off the kava. It is always hard to say goodbye on these occasions. The children were waving as we sailed past their beach on the way out of the lagoon.

We day sailed along the coast to the town of Lautoka. The anchorage there was uncomfortable, so, after a visit to the dentist and eating the scrawniest curried chicken lunch ever, we moved to Saweni Bay. There, we met Peter, a film set builder, on his Rustler 32', excitedly waiting for his surfer girlfriend, Ginny, to arrive from Auckland. Interesting and fun to be with; we still hope to see his name on the credits of films made at Pinewood Studios!

We had a rendezvous with Gaby, a friend from NZ who was coming out to sail with us for a week. While waiting, we met a father and two young sons fishing in their outboard-engined 'panga'. They had come aboard for a coffee when suddenly the father leapt back into the 'panga' while asking for a weapon to whack a sea snake that he had noticed crawling up the engine shaft! I gave him a big spanner, but too late; the snake had already curled up inside the warm engine cowl. Sea snakes are venomous and they will attack when cornered. Starting the engine was an idea, but not a real option. This was a big snake and scaly, not soft like an eel, and could have damaged the electrics when the flywheel started to turn. The fisherman knew the reptile and lifted the cowl at arm's length, and she slithered back into the sea and not onto his warm bare feet. He told us that sea snake bites are not painful, but the venom is slow acting and life threatening.

'Back to the connected world'; we anchored off the Sheraton Hotel in Momi Bay and paddled ashore to collect our artist friend and steal a jerrycan of drinking water from the beachside showers, the security guy turning a blind eye. As our visitor was a day late, Nicole decided to take the opportunity to chill out five-star style, poolside, in her best bikini. As an uninvited guest, she needed the camouflage of a towel like all the other guests. When asked for a room number, she intimated that she was from the yacht, and the attendant gave her a Sheraton towel. I decided to take my new girlfriend to the nightclub. I thought it would be cool to dance and share the rhythm of a live band. Wrong. My dancing/shuffling was deemed out of tempo and out of date, and we never went dancing again. I was feeling my age; not too good for the relationship or the ego – until that time, I thought that I was doing pretty well with my 30-year handicap! That was soon forgotten when we got back to the beach in our best clothes. There were breaking wavelets to negotiate in our soggy inflatable, ugh.

I had tasked Gaby with bringing an auto-helm spare part, duty free, from NZ. Fiji Customs thought it was a weapon and duly confiscated it. I spent the next day at the airport, haggling with officials, and had to pay import duty for the tiller arm. Although the hotel was luxurious and the staff very accommodating, we were glad to head on out to a tourist resort, Musket Cove, on Malolo Lailai island. The anchorage was crammed with yachts and float helicopters. Watching the pilots nip onto a float to grab the mooring was entertaining. Watching my two ladies shouting abuse at Nicole's ex-skipper as he arrived with a new woman was hilarious.

Gaby had not realised that, if the wind was light or from the wrong direction, it could take a long time to return to the Sheraton on Viti Levu. She had a nail-biting trip to the airport, landing in the inflatable, then racing through the lobby, wet and sandy, searching for a taxi to the airport.

We skipped Vanuatu – Nicole feared catching malaria – and we sailed for nine days towards New Caledonia. We had a tense landfall, south of Goro and Ile Ouen, as we did not have a chart of the channel through the reef. We had a depth sounder screen that stayed blank until the depth was less than 100 metres, by which time we were already over the reef, and then I was surprised by how visible the bottom was at 10 metres. The strong breeze had pushed us over the reef and into the lagoon and the instrument was still showing 10m. Ah ha! No wonder we could see the bottom so clearly; the instrument dial display was frozen at 10 metres! The water was so shallow that we nearly hit the reef and that would have been the end of Taraipo. Before going ashore in Port Moselle, the capital of New Caledonia, a boat on the next pontoon asked if we had any fresh food left, if so would we like to give it to them before Customs came down and confiscated it; we obliged, of course.

Nicole – fishing

I had heard it all: 'you do not catch fish at night' – 'you only catch fish at night' – 'sailing boats are too slow to catch fish' – 'you should not go too fast to catch fish' – 'you should make noise with tins on a line behind the boat to attract fish'.

Everybody was very willing to share their fishing success secret. It was quite amusing sometimes. However, I developed my own technique with a bicycle inner tube as a shock absorber, good wire attached to the hook and simple squid mantles as lures. All the expensive lures we bought were snatched by fish and, whatever people say – even the strongest nylon is not as good as wire.

Some places I did not catch any fish – but I was not surprised, seeing the sea lit up by fishing boats who attracted fish with light at night. It was like a science fiction film, all these bright spotlights in the dark night.

I could not count on catching a fish, but it was rare on a crossing not to catch anything.

Mahi Mahi

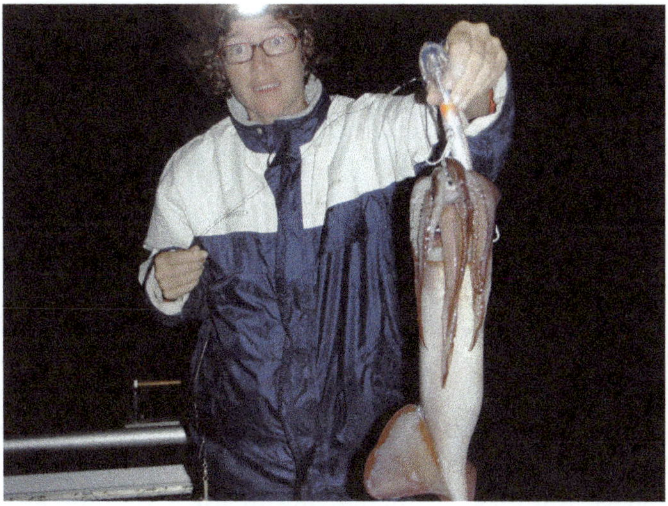

Big surprise night catch

To get away from the madding crowd we decided to go the Loyalty Islands, an overnight sail to windward. The lagoon there is a sailors' delight, a vivid white- blue -green, with an even depth of 3 metres, and no coral heads to bump into. It is about 15 miles wide, which means that an anchorage can quickly become untenable if the wind suddenly changes direction. We laid out both anchors, and went ashore to find an ATM, buy some bread and vegetables, and some 'easy-start' spray to help start our reluctant engine. Fortunately the only store in Fayaoue had some rusty old cans of 'easy-start', cheap, and after some exploration we did find a cash machine to get some French francs at the airfield.

Next day the wind did change direction, we hauled up the kedge anchor and warp ready to head to the other side of the lagoon, then the captain pressed the starter button, while the crew squirted 'easy-start' into the front cylinder air intake. It took us twenty minutes to get it running. Then followed a panic to haul up the 100 feet of anchor chain, and the 15kg anchor, as the wind and waves set in and the engine revs faltered. We desperately sailed for 2 hours, the wind started to veer, and suddenly we were free to escape towards the narrow entrance about ten miles away. The wind died and we drifted and paddled to the nearest anchorage. Amazingly on this infamous Kanak island there was a friendly black and white Breton flag flying (which brought warm memories from my time in Brittany while working for Les Glenans). I went ashore to investigate. We were in a desperate state. The engine was seized, the next time the wind did 'one eighty' we would be washed ashore. I had tried all the tricks I knew, to persuade the front cylinder to fire, including re-seating the injector. It was a nice surprise to stumble upon Joel from Brest, the only diesel mechanic on Ile Mouli. I could not believe my luck he invited me in for a Breton breakfast, crusty baguette, pork rillettes, and 'un bon coup de rouge'. Then with his precious brief case tool kit we paddled out to *Taraipo* and set to work removing the cylinder head. The bad news was that we did not have a spare head-gasket. He was due to fly to Noumea the next day. He went to the scrapyard and came back with two half gaskets from different engines with which to re-seat the ancient Renault Couach cylinder head. What a genius and it worked, relief all round as Nicole was itching to get back to civilisation in time for her best friends' wedding on the other side of the world. We romped back to the east coast of New Cal', the spectacular cliffy anchorage at Hiengiene, and found a bus to the airport next day. This was a big moment for me it seemed that Nicole was leaving with all of her belongings, sea boots, and all. She was off on a reality check. My reality was waving goodbye to the back of the airport bus, after such a whirlwind affair, a cosy night in the Youth Hostel and an emotional farewell. I could not believe that she had gone.

Jilted

So what next, rather than mope around New Cal' like a jilted lover I decided to sail to the north coast of Australia, single-handed, quite a challenging prospect with my old paper charts, no plotter and rough seas. I left in fading light to thread my way out through the reef strewn entrance. As the wind and current were pushing us sideways towards the reef, the engine rattled like a tin of nails, I switched off immediately and managed to scrape around the edge of the reef under sail only. I guessed that the coupling (a CV joint from a Subaru car) between the engine and the propeller shaft was coming apart. I had no intention, of going back, even had I been able to. I was too busy sailing clear of the reefs, and keeping a lookout for shipping, to investigate the engine. I was also carefully plotting my course, in poor visibility to clear the north end of New Cal'. Once clear next morning, I could get my head down for some shuteye. I was looking forward to an exciting month at sea, heading north-west across the Coral Sea between Australia and Papua New Guinea, through the Torres Straits and across the Arafura Sea to Darwin, approximately 2,500 miles in total.

I have learnt that I only needed approximately five hours of sleep per 24 hours; any more than that was sheer indulgence. When at sea, I can never sleep for longer than two hours at a time in daylight and I confined myself to catnapping for periods of 10 to 20 minutes at night or when there was a lot of shipping about. I napped sitting upright on the plywood bench. This was something that I used to do when taking a break at motorway carparks on long trips. In those days, I drove up to 50,000 miles per annum, often loaded down with a squad of young sailors, a boat, or windsurfers on the roof, and towing a RIB when doing my National Coach's job.

I was surprised by the relentless strength of the wind as I sailed into the funnel between Australia and Papua New Guinea. The wind was so strong I had to lie a-hull for two days and nights waiting, as I did not want to be ducking and weaving through the Torres Straits rockery in 40 knots of wind. The wait was quite worrying. I was receiving accurate strong wind forecasts on short-wave radio from 'Metal Mickey' in Australia, but had no idea when the wind speed might drop below 30 knots, and I was drifting inexorably into a dead end. The only exit could possibly have been up the Fly River never to be seen again. I had no charts of that coast. To stay on track, I gybed every 12 hours in order not to drift too far towards PNG. When on port tack, the waves crashing against the hull were brutal; this was the first opportunity for me to test the stability of the Polynesian catamaran design while lying in my bunk. I was so scared that it took me a while to pluck up the courage to take off my safety harness, waterproof jacket, trousers and boots. Though the windward hull lifted with each breaker, I never felt that Taraipo was going to capsize.

With an improving short-wave radio weather forecast from the booming automated voice of 'Metal Mickey', I hoisted the small jib and triple-reefed mainsail and headed into the Torres Straits, hoping to find shelter behind one of the low islands. There was no shelter, so I persisted in tacking back and forth across the channel. I got seven hoots from one ship for 'not making my intentions clear'; the watch keeper was obviously not aware that a sailboat couldn't go straight into the wind. When daylight came, I was in open water, with 10 miles to run before entering the channel north of Hammond Island, and I dozed off in the cockpit to be woken with a jolt by a Customs cutter alongside. The officer politely ask for my credentials. I said that I was hoping to stop at Thursday Island to access the internet to see if there was a message from my girlfriend. Sadly, the current in the channel was so strong I couldn't stop and I was ejected without ceremony into the Arafura Sea, to face another thousand miles in strange warm brown waters, and an admonishment by radio from a Customs spotter plane. They asked me for my passport number and other personal information on open Channel 16. Not very friendly, I thought. The following evening, I was blasted by the spotlight of an Australian Customs chopper; I guess my raft looked a bit too much like an Indonesian escapee. Then I had a week of peaceful drifting as far as Melville Island, where I had to choose between wriggling through the inside channel with the strong currents or being blown past Durban attempting to sail the simpler route outside the island. I did not want to risk the chance of somehow hooking up with Nicole again.

I took the inside route and spent the night 'buoy hopping' under full sail and spinnaker trying to go faster than the current in order to maintain steerage and not finish up on a mud bank in the dark on a falling tide. I exited the channel at dawn, changed course for Darwin and woke up later to find that I was being sucked backwards to where I had come from; I bore away to get more speed. An hour later, Taraipo was accosted by an enormous RIB full of guys in flak jackets and helmets. The Australian Navy no less, who asked if I was okay and if I needed anything. How nice, I thought. "My VHF is a bit weak; could you call Customs and tell them I will be there at tea time?"

"No problem, we can call them from our warship (no name)." The ship was out of sight below the horizon, and off they went, to return half an hour later to confirm that Customs had been informed, and to shower on me gifts of fresh bread, bottles of water, fruit juice, fresh apples and oranges. I guess I looked tired and hungry after being at sea for a month. Customs weren't quite as friendly when they finally got me in their clutches at Cullen Marina. They removed all my food, including the fresh stuff from their navy! They took my Visa card number, dialled it into their cell phone and removed $450 from my account before I had even set a foot ashore. It was Sunday night, so there was no chance of finding an internet café, but a friendly backpackers' place let me check my e-mail to see if there was a message from Nicole. There was one, but it was a month old. *She is not missing me,* I thought. I had to climb a security fence in the dark to get back onto the ferry landing and had to motor out and drop the anchor before the morning ferry arrived. I was determined not to lock-in and pay marina fees. The only feasible anchorage was sheltered by a beautiful high sandbank, but not at high tide, which was unsettling if the wind turned and started to drag the anchor towards the shore in the night.

Right turn

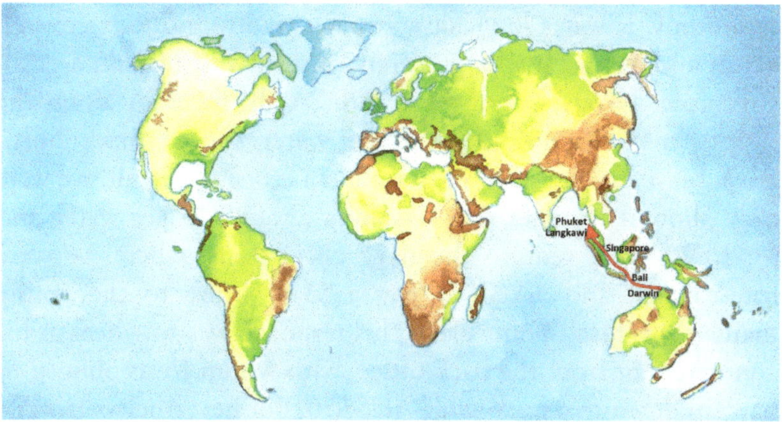

Within a few weeks, the girl was back aboard, but with a change of plan! Instead of sailing across the Indian Ocean to South Africa, we were to go via Bali, Singapore and the pirate-infested Malacca Straits for a Christmas party with friends in Langkawi. Not my choice of route, but hey, what do you have to do for love? We prepared for the long haul. I found the copy shop that could supply a complete portfolio of photocopied charts for our reroute. Surprisingly, there was a flag maker in Darwin who could supply a new red ensign and the essential courtesy flags for Indonesia, Singapore, Malaysia, Thailand, Maldives, Mauritius, Madagascar and Mozambique. You need a lot of flags to sail around the world. We were not able to able to buy the compulsory cruising permit for Indonesia from their embassy, and we delayed cleaning below the waterline for fear of seagoing crocs and poisonous transparent box jellyfish until we were in a more open anchorage. There, I donned my mask, took the plunge and cleaned off the hull, the water cooling intake and the speed impellor, while Nicole stood guard with the rolling pin!

Nine hundred and ninety miles; 18 days of calm sailing. No swimming – there were carpets of red jellyfish. Occasionally, dolphins swimming and playing around us would break the monotony. It was a long, slow trip across a turgid dust-covered ocean and against a strong current gushing out of Bali; a desolate place. It was so hot we couldn't walk barefoot along the rickety metal pontoons, and the water supply pipes burn. Because we did not have an Indonesian cruising permit, we were only allowed to enter and stay long enough to get water and food. We were sent to the navy checkpoint, where we were interrogated and asked to pay a $100 bribe to the young matelot. I refused point blank; his response was to wheel in his commandant from the back office, with whom I was obliged to bargain down to a reasonable AUS$50, but nevertheless It was a bribe, which went straight into his pocket. I was on the back foot, being deemed to have 'entered illegally'. We were sent to a nearby photocopy shop to copy our passports, etc., as they did not have a copier. On our return to the navy checkpoint and while the haggling was going on, my crew, who had been told not to say anything by way of conversation, asked the Commandant why there was a box to be ticked on the form saying that we were/were not carrying opium? I kicked her ankle as interested parties hidden from adjoining offices joined in the discussion. I envisaged the decks of our wee catamaran being trampled by the boots of a naval search party and the skipper finishing his days in a hot prison cell. When my second in command brought out her Aussie dollar coins, there was a hush as if they were pure gold; in fact, they mistook the $1 coins for $5 coins, which was fortunate for us! Bali has one of the most interesting entry and exit procedures if you like that sort of thing; it took us two days to enter and another two days to exit. Long enough for us to get a feel for the place, fill the tanks with warm water and buy food from the Makro (UK) supermarket, where the underside of every vehicle entering the car park was checked with a bomb-detector trolley. There were flowers laid in every shop doorway to protect them from bad spirits. There was nothing too spiritual about our stopover in Bali.

Nicole – time

And time becomes timeless
Day after day in the width
Of the endless sky
Shooting stars so many
That all our dreams should come true
Losing all sense of time
Surprised that another day is over

Trying to catch the green flash

Nicole – where to go?

I want to go north-west
But the wind blows from the north-west
Currents are so strong in the 'wrong' direction
Nature seems to be against me.
But it's not.
It just happens that I am there in that moment.
I can choose to go south/north instead
Not losing my goal
Not getting desperate that my goal doesn't get any closer
And trying to do the best I can.

Nicole – calms

Calms that make you doubt that you will ever arrive. Nevertheless, we are advancing slowly, doing 40 miles a day, but only three in the right direction, sometimes two back. Never losing the confidence that we'll arrive with the help of the wind, using the engine when nothing else is possible.

Wind is endless; diesel isn't.

You could motor the whole way, but then you'd never hear the silence, the waves softly turning around the hull. You wouldn't hear your boat when she decides to go forward. Hear the boat – is that possible? It's only wood! But maybe that moment you were connected, being on a boat, one hand in the sky and the other one in the water – feeling that times are changing. The calms are over and it's going to be fast!

After 24 days and 1,180 miles, we arrive in Changi!

We had a slow and windless passage to Singapore, seeing the same islands for days on end, dodging amongst fast-moving vessels in the oilfields at night and having to sail past Borneo without stopping to see the apes, as we had to meet up with Nicole's friends for Christmas in Malaysia.

Singapore hove into view as we approached the towering skyline on a light hazy morning, but the route was barred by two lanes of moving supertankers as far as the eye could see. For us to cross both lanes without a collision, we needed speed, for which we needed wind. We bided our time. A squall appeared over the skyscrapers; we went for it, then white-out – we couldn't see a thing in the stinging rain. Then out of the murk appeared a police boat to stop us in our tracks. As the rain cleared, the wind was dying as we panicked our way across leviathan bows to port and starboard. There is no anchorage suitable for a small vessel amid the hundreds of ships waiting their turn to load or unload. We reluctantly paid to enjoy the luxury of a swinging mooring plus water taxi at Changi Sailing Club. No worries about anchors dragging through the soft mud and strong currents; we could spend worry free days ashore enjoying the contrast of life in a vibrant, air conditioned, highly disciplined, chewing gum free city state.

It took us half a day of bureaucracy to escape from Changi, because Nicole's passport had carelessly not been stamped on entry and logically could not be stamped on exit, the officer argued. If I had not summoned his chief, I think we would still be there.

Pirate country?

The infamous pirate-ridden Malacca Straits were the next challenge. There are so many big ships passing through the straits that they are obliged by International law to stay within a defined track each way. Outside of these tracks is a free-for-all, where we found ourselves, night and day, ducking and weaving among trawlers, high-speed ferries, tugs with long tows and mile-long fishing nets stretching out from the shore. To add to the frisson, they could all see each other on their radar screens; equipment we did not have. The interpretation of the ships' lighting regulations in the third world left a lot to be desired. Tugs often did not light their long tows, and fishing boats rarely showed port and starboard lights, often showing flashing red green or amber lights; ships now have arc lights at the stern to ward off pirates – all very confusing. I never thought for a moment that our tiny little raft would be a target of the pirates of Malacca.

Halfway up the Straits, we decided to stop for a rest inside the sandbanks of the Muar River. "What is that? It looks like a marina with no yachts in it." It was a brand-new marina awaiting its official opening ceremony. Unfortunately, on the ebb, there was about five knots of current running through the pontoons, which meant that we didn't finish up on the one we aimed for. Not a problem when the place is empty and the fenders are out. The marina official showed us around the 'multiple-purpose function rooms' and then insisted that we remove our vessel from the marina, as the opening ceremony was still to be held! So we had to anchor just outside the marina in the roaring current, in the middle of Muar, an intriguing city with mosques either side of the river; we couldn't wait to get ashore. Lunch was a delight. Here, we learnt how to eat without a knife and fork, 'left hand only', and use the sinks provided to wash our hands between courses; quite a ritual, but the food was great despite the fact that we never understood what we were ordering. We were made very welcome, but we drew a blank on day one in our mission to find a sailmaker to repair the tired mainsail that we had torn from leech to luff in a rain squall. The next day, we found a Chinese man with an industrial sewing machine, who, to our dismay, dragged our lovely white sail across the oily floor as he rushed it through his machine. In the process, the sail lost some of its shape. He normally repaired truck tarpaulins, but 'hey, we can sail again', which was just as well, as the engine was virtually dead; it was only firing on one cylinder and we didn't have time to hunt for a mechanic as we were on a mission to meet our friends for Christmas in Langkawi. Oh, the stress of it all!

Nicole – pain and agony

'I get on a bus' – 'next stop, I'll fly home' (wherever home might be for a live-aboard person...). Everybody has way of expressing the rough, uncomfortable times on the ocean when one wants to be anywhere else than on this particular sailing boat, with this particular bad, windy weather, with that particular person. Dreaming of a 'normal' life, a firm roof over your head; nothing moving but rock solid; an evening out with your dearest friends. Even having a boring, but regular job seems to be attractive in that moment.

You simply don't want to be shaken around again; you don't want the squall to come when you are on watch, the autopilot to stop in the middle of the night, the sail to tear from luff to leech, the engine to break down. You don't want to calculate how many apples there are left, but to go in the supermarket and buy some.

You'd like to have another person to talk to, especially your best girlfriend; you'd like to see your godson growing up, to eat vegetables grown in your parents' garden. Snow seems to be a dream – finally cold, skiing, hot baths and coffees in nice coffee shops.

For me – how bad the weather, how often the squalls, how little sleep I have – the most uncomfortable times are when I miss what I don't have.

We were running short of ideas as we approached the coast in Langkawi in the dead of night; we were seeing myriad small white lights low on the water. We were tired. "Are they fish traps?" We tacked quietly in until we made out that they were small fishing boats, each with a snoring fisherman aboard. We continued into the shallows, anchored and waited for dawn. As daylight broke, they all roared off with their fishtail engines, leaving us in absolute calm, to be broken a few hours later by a wave splashing through the port light onto the chart table and my bunk. I leapt up to see an idiot on a jet-ski tearing around us, showing off. I wished I had a deck hose to attack him with a water jet. We made our final approach through the boats in the main anchorage with some panache; it was quite windy and the engine was dead, much to the surprise of the cruisers, one of whom rowed over to give Nicole a hand. We anchored under sail, close to the beach.

Nicole's recipes

Aku Ankka's rice wine

1kg rice
2.6kg sugar
2 handfuls raisins
2 lemons (thinly sliced)
10 litres water
2 tsp yeast
1 cup strong tea

Dissolve sugar in 2 litres of boiling water.
Into a clean (sterilised) white bucket, add washed rice, 8 litres of water and the rest of the ingredients.
Stir after 20 min, and then daily.
Cover opening with a cloth to avoid fruit flies.
Fermentation stops in about 2 weeks; strain into another (sterilised) bucket and let it settle for 1–2 days.
Siphon into sterilised bottles.
A week later, decant for clear wine.
Enjoy!

The engineer we had called was waiting on the beach barefoot and holding a small canvas tool bag. Ar Kumar was coming to resuscitate the engine.

Ar Kumar

Sitting with his feet in the oily bilge, he removed the cylinder head, the injectors and injection pump and took them ashore to fix them. Two weeks later, as promised, he reassembled all the bits and tried the starter. It fired, but there was no cooling water flowing out of the exhaust.

Oh dear, it was nearly Christmas and it was dusk; Ar Kumar was deep in thought. He asked, "Do you have an electric drill and a 5mm stainless bolt?" I brought out the little Honda generator, the drill, and a box of nuts and bolts. He put his finger on the engine where he thought that dried salt must be choking the channel inside the cylinder block. He carefully drilled a 5mm hole, then requested a stiff piece of wire that we didn't have. He rummaged through the galley, found the whisk, cut off one wire and used that to poke out the salt deposit. We didn't have a tap and die set. Undeterred, he proceeded to insert the bolt into the aluminium cylinder block with brute force, and bingo, it all worked. We paid him $250 and he disappeared into the tropical night, and we would be able to rendezvous with our friends for Christmas decreed. We anchored off the Sheraton Hotel and paddled ashore in our 'deflatable' dinghy, to find ourselves shunned by the staff, we were obviously not residents 'we were too suntanned' and our dinghy spoiled the tidiness of their beach. Fortunately hotel resident friends Conny and Beat were there to negotiate the use of deckchairs and the pool bar. Not that I can lie on a deck chair reading for more than an hour, or drink many beers in the tropical heat of the day. Thailand was on Nicole's bucket list so that, was the way we had to go!

We were sad to leave friendly Malaysia, but excited at the prospect of visiting Thailand. Preceded by an overnight sail, the night was as bright as day as we drifted through myriad arc lit fishing boats, draining the ocean of every remaining tiddler it seemed. The port of Chalong in Phuket was not as I had imagined it would be. Of course paradise is never quite as imagined. Landing at the yacht club (was not a club, it was a bar) was fronted by deep mud at low tide. Leaving at low tide required a volunteer to push the dinghy, sledge like, knee deep in the goo.

We were there post tsunami every tripper boat had two brand new 200hp outboard engines on the transom. There was nowhere for us to hide from the wash and the noise. One boat even slashed off our trolling fishing line, which wrapped around his propeller and stopped the engine dead, much to the surprise of the dozen or so day trippers aboard. We found it difficult to get ourselves heard in such a frenetically tourist venue and headed out into the Andaman.

Nicole - Now

The longer I sail the more I discover what is important to me, which isn't necessarily what is important to others. I like the slow advancing, only hearing the birds, the waves slopping against the hull, John's breathing while he's sleeping. Mike, our autopilot's, little movements.

I prefer silence to speed, wood to high-tech for the trip I am on now. I am sure that another time will come and I'll enjoy a fast boat.

But, now is now.

For one happy moment (on a weeks' trip) like this it is well worth going through doubts and rough, uncomfortable times.

Sailors Shangri-la

A long slow trip across the Indian Ocean was the best antidote to Thailand. Addu Atoll, in the Maldives was the destination, peaceful and friendly, though like most atolls short on shallow water for anchoring. We were limited by the power of the crews biceps as to how deep we anchored as there was no windlass aboard *Taraipo*. More than ten metres deep and the anchor and chain hanging vertically became a serious challenge, requiring lots of ingenuity, chain hooks, stoppers, and the sheet winches to recover the anchor on a windy day. After a rough night on anchor watch, we discovered the tiny creek sheltered by the causeway between Gan and Feydhoo. Bliss a cosy anchorage shared with lots of boats, and serviced by the friendliest locals possible. We learnt how to chew betel nuts (not nice, not healthy, and your teeth go red) and how to peel a coconut having failed previously with a hacksaw. We did not stay any longer than necessary 'we were on a mission', determined to get to Chagos the 'sailors Shangri-la'.

I never missed an opportunity to practice our sailing skills, breaking out the anchor and peacefully negotiating the narrow pass without engine, watched by a knowledgeable audience of sailors gave me a buzz.

As we left Addu we were loaded down to the waterline with enough food and water to sustain ourselves for a month or two in the deserted Chagos Islands. Ten miles out, we spotted the local tuna boats in action; exciting stuff. They have a unique fishing technique. The boats are about 20m long, high powered, and compete with each other while chasing the tuna shoals. When they are 'in amongst them', one man each side forward pours 'rubby-dubby' (bait) over the side, while 10 men are on the wide deck aft, each armed with a 2m pole and short line tipped with a barbless hook. They hook 5–10 fish per minute, tossing them over their shoulders directly into the hold! The frenzy created by compressed air bubbles from a pipe aft lasts for about 15 minutes, after which the boats search for another shoal.

Nicole's diary

Eight days after leaving the Maldives, wonderful anchorage; 320 miles on the ocean.

What does it matter that we take three times as long as other boats? We have a wonderful sail, an unforgettable journey through, over the ocean. If I only wanted to see places, it would be far easier to catch a plane. It's the journey itself that is in the centre, focused on a goal we might achieve. Isn't it like life itself? Waiting and longing for something doesn't help. Life is a journey, not an arrival. Let's have as great as possible a life, with squalls and calms, making decisions that lead us to where we want to go – a fulfilled life – and lots of hurdles to be jumped over, successfully.

The journey itself; that's it. Jealousy only comes if we compare ourselves with others – we are not fast enough, our boat isn't big enough, our sails are not strong enough. But we can handle our boat; we get where we want to go with our own possibilities, with the decisions we've made. We decided not to spend money on a new sail as long as this one lasts. We take care of it.

Let's take care of our lives, of our journey, not wanting too much, but making wise decisions that lead us to smile and the satisfaction at the end of it. Being happy with what we have. I could go on a million-dollar boat – but would I have moments like this, going slow and having time to write? Or would it be too fast, the night disturbed by engine noise?

We only had second-hand charts and scant information about the Chagos Islands. As we approached Ile Diamant, we were nonplussed to have the masthead VHF aerial attacked by a vicious Booby. We had heard that the archipelago was deserted, so it came as a shock when a speed boat appeared from over the horizon. 'Stand by to repel borders'; it was a navy RIB commanded by a handsome young officer who explained that we were in the British Indian Ocean territory, something that most Brits didn't know about. It is a strictly controlled marine nature reserve, and we had to pay $100 to stay and had our passports stamped. I was embarrassed to hear that the British had previously evicted the whole Chagos population to Mauritius at the behest of the Americans, who wanted to hire Diego Garcia as a strategic naval base in the middle of the Indian Ocean. We visiting yachts were all corralled into one anchorage when a cruise ship hired by protesting islanders came to visit. The younger generation wish to establish the right to return. I think that they should have that right. Unfortunately, they probably no longer have the skills to survive in such an alien environment.

Forbidden fruit!

Hauling the anchor

Our 'raft' was adopted by a shoal of squid, and crews from other boats came aboard Taraipo for fishing parties. Squids are smart; they hide under the hulls and only dart out after our lures the instant they splash into the water. After a few days, they understood what was happening to their population numbers and deserted us. When Michel, a solo sailor, arrived on his annual pilgrimage from Thailand, he went hunting big land crabs and harvesting oranges from trees that he had secretly planted in previous years. Boddam atoll, the only safe anchorage, attracted more than 20 yachts that survived the summer there, sharing the solitude, their stocks, barbecues and games in clearings ashore. The French crews run the weekly boules tournament, and the Americans the beach volleyball games. As time passed, all this intermingling can challenge relationships; couples that have been together all their working lives can find living in the same space 24/7 quite a challenge. Many of the boats made permanent moorings by wrapping chain around 'bommies', and the smaller vessels soon run out of booze and have to resort making rice wine, which can be reasonably palatable and is certainly alcoholic. Being in the middle of a marine nature reserve, we all caught and ate the most amazing fish.

Nicole's recipes

Conserve de poisson – recette de Sylvie, Tom and Philippe

Mettre du poisson cru dans des bocaux.

On peut mettre un peu d'huile ou d'eau de mer dans les bocaux, mais ce n'est pas obligatoire car le poisson frais rend beaucoup d'eau. Mettre les bocaux fermés dans la cocotte-minute (beaucoup d'eau dans la cocotte), laisser cuire 1 h 10 les petits bocaux à partir du moment que sa siffle. (1 h 30 si grand bocaux).

PS on peut mettre du thym ou du laurier.

Because there is no commercial fishing, there are a lot of sharks in the lagoon; it is my belief that they do not like human flesh, but they are scary beasts. I was always careful to keep my toes out of the water when paddling ashore on the windsurfer, especially near the old village jetty where sailors cleaned fish.

Nicole – enjoying life

April – 12 Mai 2006

A freshwater bubble bath after a squall

Diving for shells
Catching squids and fish
Drinking fresh green coconut juice
Swapping recipes
Rowing around the anchorage
Collecting shells at the beach
Playing games on Moonlighter
Drinking home-made wine with 'Asi Sea'
Sleeping outside
Baking bread on the fire
Having coffees with 'Shayile'
A carrot from Daniela!

Catching a crab
Playing cards on Yellow
Rowing over to Belagaer
Cutting Jean's hair
Cooking fresh fish in fresh coconut milk
Sailing Jack and Sandy's dinghy

A quiet moment in the shade

Writing poems
Meeting Norbert and Andrea
Drinks on Taraipo and other boats
Playing volleyball
Having a party – cinco de Mayo
Making jewellery with Anne
Sailing Tuggii with umbrella
Listening to the concert from the next boat
Evening with Bernadette and Roger
Reading books
Having a coffee in the morning on the trampoline
Being happy with myself

Africa, here we come!

We were fearful of grounding; with Diego Garcia below the horizon we were traversing large tracts of shallow water in the middle of the Indian Ocean on the way southwards to Rodrigues Island. We were the first arrivals of the season. Raj, the immigration officer, suggested that, before formalities, we should rush to the Saturday market, which closed at midday; cool. Later, he invited us to his house for a cup of tea. Raj's day job was chief Medical Officer for Mauritius, and he was on secondment to Rodrigues for six months, 500 miles away from his family. He warmed to us when he discovered that Nicole could beat him at Scrabble despite his loose interpretation of the rules of the game. The main health problem at the time on the island was a programme of mosquito eradication by teams with chemical fogging guns. *En passant*, Raj mentioned how good the free national health system at the island hospital was. If we had any health issues, he could introduce us to the Director. Nicole had a problem with her toenails, and I remembered that I had a bit of a lump in my tummy. The next morning, Raj drove us up the hill in his official car to the hospital. A potion for the toenails wasn't a problem, but the surgeon reckoned that I should consider surgery either there and then, or at the main hospital on Mauritius, to where he would be returning shortly. Having absolute confidence in the ability of surgeons everywhere, and a protected lagoon to park Nicole and Taraipo in, I signed the disclaimer form. The next day, Raj and Nicole were delivering me in style to the hospital, until he realised that I did not have a knife, fork and spoon, a plate, a blanket and a pillow. I lay on the bare mattress in the busy men's surgical ward, a large fan overhead, and hoping that I was going to survive the operation for a hernia, on an old wound. At lunchtime, family members came in to feed and look after their patient, and me! My recovery after the op was so fast that I was able to watch the football World Cup on TV that night with the duty nurse, to the accompaniment of the moans and groans of some very sick men. The next morning, I signed myself out and walked gingerly down to the harbour to see my girlfriend riding around town with a stranger! 'No harm done'; they carefully got me into the dinghy and back aboard. "Phew, I've survived another dice with death, and glad to be back afloat." I had strict instructions from the surgeon not to do any heavy lifting for three months, which meant that Nicole would have to 'take the strain'. She was invited by a local skipper to a day out aboard the Blue Dynamite fishing for tuna and sharks. She couldn't resist, but did have regrets halfway through the day when she realised that there was no pee bucket provided, and that the crew believed that it was bad luck to have a woman aboard! From my invalid's seat, through binoculars, I watched in disbelief as they chugged back in past Taraipo with a massive tiger shark, the length of Blue Diamond, tied alongside. It was the biggest that they had ever caught and too big to be craned out on the jetty; it had to be dragged to the beach and carefully carved up. 'Nicole smashes yet another fishing superstition', and made me proud by association! The technique that they used for catching shark was to anchor bait overnight and collect the dead sharks the following day. The cyclone season was nigh; we had to head for Mauritius ASAP. The wind shadow behind the island lulled us into leaving too soon in what turned out to be half a gale of wind; once at sea, there was no chance of turning back, 'stitches or no stitches in my gut'. Nicole dropped the mainsail and we careered downwind with the smallest jib and 'Mike' the autopilot in command.

On approach to Port Louis, we were at 'ramming stations' with a departing tanker when we realised that our VHF was expiring. On arrival at the Customs wall, an officer bellowed down, "You are not allowed to come in here without calling on VHF." We replied, "But we are here; our radio is broken."

"Well, you are against the regulations." At which point, he directed us alongside the highest wall rather than the visitors' pontoon. He took our warps and told us to report to the Customs office with the ships papers, knowing that we would need to be wearing crampons to climb the wall. Surprisingly, the 'freedom of the seas' included the freedom to visit any country in the world, and usually to be allowed to stay for a maximum of six months (we didn't try North Korea). After doing Customs and Immigration, we moved to a friendly dock. Raj and family came by for a picnic and we set about the serious business of finding a way of removing the cracked fore-beam that supported the mast. A very plausible young man offered to haul Taraipo up a beach somewhere in the backwoods with a tractor, dismast and replace the beam 'for a very good price, my friend'. We saw him later downtown and he tried to tap me for an advance to buy hash; a casino bouncer saw what was happening and discretely warned me off him. The next day, I found a boat builder, who showed me an immaculate yacht interior that he had built. He was obviously our man. He turned up on Saturday morning with a truck full of pals and took our boat to bits; very scary. They removed the bad fore-beam and took it away to model a new one. This operation took place in the middle of the fashion shops and tourist hotels of Port Louis, without a complaint. The gang was back the next morning, having worked most of the night laminating an exact copy of the old fore-beam, plus four new wooden hatches. Job done, precisely, and to the agreed price. He also found me a supplier of cheap antifouling paint for our next haul out. Raj and Indira invited us for afternoon tea and some fast driving around the sights. He had been a rally driver in his youth and his wife the co-pilot! We could have stayed longer, but Nicole was on a mission to go to the Seychelles, a lifelong ambition, despite the fact that this folly would entail a long sail to windward. After one night of roughing it, she capitulated and we pored over the chart looking for an alternative high-end venue, and 'lo and behold', Mayotte was beckoning on the far side of Madagascar. We did not want to stop in Madagascar for fear of dengue fever. Mayotte we had never heard of, but it looked nice on the chart. Passing the northern end of Madagascar would be very windy, and so it proved to be; we tore down the long swells at 10 knots, leaving two white trails behind us like a sledge in the snow. The wind then petered out. Mayotte is in a fickle windless zone in the middle of the Mozambique Channel, but well worth the slow drift to get there. It is a magnificent volcanic island encircled by a coral reef.

After the long approach, we found a mooring in the village of Dzaoudzi. When we got ashore, we were shocked to see that the African women had painted their faces yellow; everyone spoke French, but were in fancy dress! In fact, their make-up was a paste made from crushed coral to protect the skin. Mystery solved. The locals had recently voted to remain French and separate from the other Comoros Islands. This had created a nightly 'under the radar' passage of small boats smuggling poor refugees into Mayotte. Most of them were rounded up by the Gendarmes and flown back on a daily prison plane! The French expats (doctors, dentists, government officials, Customs officers, ferry captains and police, etc.) were all on generous 1.5 x French salaries and never stopped moaning about the locals and couldn't wait to return to the 'Hexagone'. We were charmed and fascinated by the mix of people we met on the ferry to the shops in the capital on the main island.

Many yachts passed through Mayotte and there was a buzzing sailing community of locals and visitors, with plentiful sailing supplies and servicing. When I was in New Zealand, I was pals with Hans, a brilliant sailor aboard a 70-foot wooden catamaran. In Mayotte, I met his dad. In fact, the old boy nearly stole my girlfriend from me one evening at a yacht club hop. The fact is that I could never put one foot in front of the other and Ernst was a classy ballroom dancer; he swept Nicole off her feet, while I was unable to persuade any of the French 'madams' to dance with me. I was not unhappy to see him sail away the next day in his leaky old black wooden boat, which he had sold for a dollar to a guy in Thailand. Taraipo was not in need of any major works, which gave us time to explore the deep lagoon (lots of nesting turtles), help Roger and Bernadette to rig a radar dome up their mast, and to do some dog minding for John and Julia, whose superyacht was so big that they had to 'anchor off' in the middle of the bay.

We were told several times that it was silly to have missed out on Madagascar. That 'Nicole is right', there is a risk of catching malaria, but pills are freely available in the unlikely event of infection. We agreed that visiting Madagascar was a must.

We renewed our 'mozzy' nets and backtracked 200 miles to Hell-Ville on Nosy Be island, Madagascar, and anchored near to the ferry pier. There was one other yacht in the anchorage, which had been impounded and the owner jailed for infidelity with his wife, a local woman. Chillingly, another owner was attacked and his headless body left sitting in a deck chair in front of the mast; 'no wonder the place is called Hell-Ville'. We were quite keen to check in and check out quickly with Customs, and leave for another idyllic tropical anchorage. Not so easy; 'step one', there is nowhere safe to leave the leaky inflatable dinghy on the only public slipway. We had to pay a local boy to take care of it; he did and he re-inflated it by mouth on our delayed return. I had somehow upset the authorities by not going to the four offices (Customs, Immigration, harbour master and government) in the correct order and I was threatened with jail for disrespectfully jumping the queue and not having the correct number of stamps on my entry papers. I was careful to anchor in the middle of the other visiting yachts that night (safety in numbers). Most of the transportation on the east coast of Madagascar is under sail, no fuel, no engines and, most importantly, a fresh wind blows offshore every morning and onshore every afternoon. Pirogues carrying people and goods sail in and out of the harbour night and day. Fascinating.

Anchored off the beach on Nosy Ambariavato, I could not believe that people had to spend their lives chipping at a 10-metre-high boulder with hammer and cold chisel to produce a six-metre pile of road chippings. A barge sailed in at high tide and beached next to the pile, and at low tide the village loaded the chippings onto the barge, which re-floated at high tide and sailed away the next day. The people of Madagascar were desperately poor, but they did not bear us a grudge and were keen to chat and sell us stuff. Even some of the poorest had mobile phones and they would paddle out to us to beg for a charge!

Great sailors; great fishermen

Mahajanga was something else. How do you bring a steer to market in a sailing boat? Answer, lying on its side with its feet tied together and a long tether hanging from its neck. When the boat beached in the city, the crowd scattered as the steer was rolled over the side and its feet released, and off it charged into the crowd to be caught and led to the abattoir! Anchored in the shallow, sandy, swirling estuary, at dawn, I counted more than 400 fishing pirogues sailing out empty with the land breeze, and returning elated with the sea breeze at dusk. Despite the poverty and hardship, these fishermen were getting a buzz from trapezing their outriggers.

Mahajanga rush hour

Earning a living the hard way

We were enchanted and reluctant to sail away; however, needs must – the cyclone season was upon us. On the first night out, we were run down by a big pirogue. It was just after midnight, a moonless night. I sensed a black mass to port, and then there was a massive impact. I ran forward to see what it was, but the rogue ship was gone. I called out to Nicole, who had been asleep in the starboard bow, "Are you okay?"

"Yes, but I can't find my glasses."

There was a lot of damage. The port stem was broken off, the anchor was crumpled, the starboard stem was damaged and the trampolines; our foredeck went soggy, as did the rig. We decided to go back to assess the damage and make repairs. There was nobody out there to help us. I suspect that we had been hit by one of the big fishing pirogues, and that the crew were dozing, counterbalancing on the outrigger and the helmsman was looking the other way. Immediately after the crash, I could hear shouting and see a flashlight in the distance, then they were gone. We headed upwind back to a likely looking beach at the entrance to Mahajanga.

Broken-off port bow

We were lucky to have beached at high water on a big spring tide in a small fishing village. The villagers were shy, but the kids soon adopted the shaded area under Taraipo's deck as their playroom. The returning fishermen came to inspect our damage. The men were embarrassed that their compatriots had not stopped to help us. They willingly helped source timber, tools and skilled helpers to rebuild the front of our home. The corrupt Immigration officer was annoyed by our return and gave us a two-week deadline to leave the country, which we managed with the help of our new friends. The fact that we are both fluent French speakers enabled us to make friends in the many French protectorates as we sailed around the equator.

Retrieving stone anchor

Kids aboard invasion; we loved them

Adze to shape the stem

Repairing the port bow

Crossing the Mozambique Channel was tricky, because the wind was so fickle and the current unpredictable, which meant that, incredibly, we missed South Africa and landed up in Mozambique! The approach to Maxixe is across a dangerous sandy bank into Inhambane. We needed high sun, high tide and a beam wind to avoid certain destruction in the massive breakers. A high level of concentration is also needed, the crew at the chart table and the skipper scanning ahead in order to keep to the dark blue bits of sea that are difficult to pick out from the moving deck at sea level. There is no question of climbing the rigging to get a better view. Good communication and fast reactions at the helm are far more important.

We landed on the end of the ferry pier, lashed the inflatable dinghy to the handrails and wandered off past the guard at the security gate to find the Customs and Immigration offices. On our return, we found our dinghy locked to the gate, having been carried the length of the pier by the guard. In front of his desk in the guardhouse, despite his explanation in Portuguese (a language we love to hear, but can't understand), it was clear that he was looking for a fine (bribe) of $100. As on principal we do not pay bribes, we returned to the Immigration office to ask for help with translation and resolution of the problem, which worked, as we required the dinghy to cross the river to Maxixe to get our passports stamped.

The next big hurdle was to find our way around the Cape of Good Hope, down the formidable east coast, which has a strong south-flowing current, frequent gales and very few refuges. New VHF acquaintances aboard Deusa, the only other yacht in the area, introduced us 'on air' to weather guru, Fred, in Durban. In order to avoid another day of incomprehensible bureaucracy and expense, we decided to leave without checking out, with the intention of skipping a stop in the capital, Maputo. To accomplish this, we awaited a four-day gale-free weather window forecast from Fred. This would enable us to reach Richards Bay on the four-knot Agulhas current, with moderate following winds. It was a long, hot, slow trip, but great to be accompanied by twice-daily personalised forecasts from Fred. Unfortunately, we were not able to return the compliment, as we did not have a short-wave radio transmitter. We were tempted on several occasions to buy a transmitter, but decided against it, as they are cumbersome, complex, heavy and require a lot of electricity, and we would have to sit the exam for an operator licence. The VHF radio could transmit up to 25 miles, and we carried an EPIRB, an emergency locator beacon, our backstop if we were in imminent danger of dying. Once triggered, it can't be stopped and could transmit thousands of miles via satellite to coastguard rescue centres, in turn alerting nearby ships had we been in dire straits. We never were in dire straits, though it sometimes felt like we were going to die.

Nicole's diary

"On the way again from Madagascar to Richards Bay, South Africa."

1 *8.11.06 close-hauled; water under my bunk. I hate it!*

2 *John repairs the leaks; hardly any water. 'Relief'; we eat the 8kg fish I caught last night. Caught another fish, 12kg; 'looks strange', all grey flesh. We decide not to eat it. It's good that John and I agree on such matters.*

3 *"Very pleasant sailing." Dolphins around the boat at lunchtime, but they don't want a picture to be taken. I finished reading* Tara Road *(for the* 5th *time...).*

4 *Bad news: we tore the spinnaker. Good news: after a hot, sweaty morning without wind, the afternoon was windy; we saw a ray fish and first raindrops in months; lovely phosphorescence in the water. "Still 992 miles to go." We are very slow...*

5 *Made necklace and cut John's hair. Close-hauled again...*

6 *Calm, calm day close to Juan de Nova; 10am, another fish; 4kg; just right!*

7 *What luxury, fish for breakfast, and lunch a nice swim, and the spinnaker up, then stitching! Once again – this time, the jib.*

8 *Hot, hot! Sleeping all day and finishing the red necklace.*

9 *Lots of movement; makes me incredibly tired. We had our last tin of apple compote with custard. I love the song* Kein Schoener Land.

10 *I couldn't sleep in the night; too windy and choppy sea. Stitched bimini the whole day long – looks nice! It was very calm, with a dolphin parade at sunset to the music* Sailing to Philadelphia. *A day full of contrasts; 680 miles to go.*

11 *Made another necklace and stitching again. It works again, the spi!*

12 *"Happy anniversary!" Flat calm, 0.1 knots at lunchtime and now 6.0kt!*

13 Survival day – 116 miles in one day.
14 Calm again; in the current doing 2kt without any wind. We plan to stop in Inhambane.
15 22.11: we're in Inhambane, Mozambique.

Five days in Mozambique

1 27 November, 'we're on our way again'. After a wonderful time with Robert and Rosemary on Deusa and a hard time with the 'jetty'.
2 Bumpy, no current and very slow course to Maputo.
3 Dolphins like the waves – I survive... course on Richards Bay.
4 Huge waves behind us – scary! Going downwind, will we get to Richards Bay? Best dolphin parade ever!
5 'We're smart' – but tacking our way. Frustrating for me. John takes it very philosophically: "Well, if we don't get there, we sit here and drink gin and tonic."
6 6am: No wind; no engine; we are going backwards, going where one of the strongest currents should be with us! I think a boat is one of the only places you are seriously 'caught'; how to unload your frustration and fast-building aggression?
- Shout at your captain – deteriorates the atmosphere;
- Running away – includes a 50-mile swim ashore;
- Clean your cabin – makes you slightly seasick;
- Breathe deep and think positive – doesn't work at all;
- A punch ball would help – but we don't have one.

I chose to get seasick and clean my cabin.

7pm: we caught a fish: Mahi-Mahi, 8kg.
10pm: four miles to go to Richards Bay!

3 December 2006; 2am ARRIVED

In Richards Bay, we moored alongside the wall inside the tugboat berth, from where we could take a minibus into town to check in with Customs, except that 'whites don't travel on buses'. Welcome to apartheid reversed; so we took a taxi like all of the other yachties and went on a spending spree buying food and stuff that we had not been able to buy since leaving Madagascar.

Rounding the Cape of Good Hope

We were befriended by a platinum miner, who gave Nicole a lift to Pretoria to catch 'the silver bird' to Switzerland for yet another reality check. I had a cool time by myself again. The harbour was buzzing with lots of well-travelled sailors to share know-how and to party with over Christmas and New Year before attempting to move on to Durban, an overnight trip for which I would need to borrow a crew in order to negotiate the tricky exit and a tricky entry in two busy ports under very strict surveillance by harbour control.

Durban marina at Point Yacht Club was going to be an expensive waiting place, so I opted to anchor in the shallows between it and the tug channel. I just loved watching the big ship gyrations, close to, night and day with no risk of being run down. Shortly after arriving, I met Fred, the weather forecaster, and wife Eva, who ferried visiting sailors to the shops in their Land Rover. They offered to collect the love of my life from the airport in the night; a great privilege, though it felt a bit like being chaperoned by the parents.

Nicole was back and things started to happen again. One day, we were going ashore shopping and I noticed a tiny wreck of a dinghy, holed and full of mud, lying on the marina pontoon. It had been dredged up by divers and the harbour master was keen for us to take it away. We loaded it onto the front of our inflatable and paddled out to Taraipo, and spent a week cleaning and repairing the thin fibreglass shell. A nice coat of white polyurethane, varnish on the woodwork, new rowlocks, oars, a sculling notch and a glued fender strip donated by Dave and Shauna, local Polynesian catamaran sailors. Now Nicole had her own tender, light enough for her to lift on and off the deck. I found sailmaker, Harry Ellens, who was prepared to make a new mainsail and jib to my specifications, and cheap too. We moved to Fred's friendly Buff Yacht Club beach for a change of scene while awaiting delivery of the sails. Harry came aboard and we bent the sails on together. They set perfectly despite the fully battened mainsail hoisting on mast hoops rather than on a track, and gained us an extra knot of boat speed.

We waited weeks for the right weather window. Early one morning, Fred and Eva arrived and whisked us off in their Land Rover to Immigration to check out in order to rush off to sea. The problem in Durban is that the wind suddenly switches from being a strong south-westerly, which creates waves in the Agulhas current big enough to break a tanker, to a strong north-easterly, which would stop us dead, in the north-east-facing exit channel. We managed to exit, but, before we could reach the Agulhas current, which was predicted to be about five miles offshore, the wind went light. Fred's midday forecast finished with: "John and Nicole, if you are listening, you would be better returning to shore." Not having an engine extended our predicted passage time. Very frustrating, as we had to report in again to Immigration, before finally escaping for the short leg to East London for a one-night stand with Norbert on Pamela. Norbert was the only sailor we met who was consistently good at catching fish everywhere he went.

Port Elizabeth is a big commercial harbour and the Algoa Bay Yacht Club looked on the chart to be beguilingly well protected. My mistake; the Indian Ocean swell finds its way past the two breakwaters. It was late in the day when we found a vacant berth on the heavy steel pontoons and were surprised to be woken in the middle of the night by violent snatching on the warps. I could hardly stand on the pontoon so violent was the motion. I doubled up everything and added an extra spring, which didn't help. Taraipo was being torn apart, breaking cleats, as she tried to charge back and forth on the incoming swell. We had to find another place to moor. At daybreak, I saw that there was one empty fore and aft mooring between two buoys closer to the shore. We dropped back onto it and waited for complaints from the harbour master; they never came. We were now committed to epic trips ashore with our wee dinghy via the charging pontoon and walkways. We regained calm aboard and respite to repair the damage.

Once ashore, we had to negotiate heavy handed dock security, as every vehicle was stopped and searched 24/7, until our taxi driver explained with a flashing smile that he used a different name that was carefully logged. Every time he drove in, he signed in 'Mr D. Duck' and signed out 'Mr M. Mouse' (they didn't notice our gas tank in a backpack). While biding our time waiting for a weather window to negotiate the fearsome 'Cape', we hired a car and explored the Garden Route inland. Went to see *Love Actually* at the cinema, and had fun times interacting with the locals around town on the buses, also with the white part of the population in the yacht club.

At sea again with only a small weather window, we opted for a stopover in Knysna, an intriguing harbour, the entrance of which can only be negotiated through the breakers guided by the coastguard on VHF radio. Fortunately, there was a change in Fred's midday forecast. He announced that we should skip Knysna and head towards Cape Town. The next day, we were becalmed off the Cape of Good Hope! There were seals floating on their backs flapping their flippers in the air, like a scene from *Water World*, but Fred was warning of violent winds the following day. It seemed to me that we could safely continue to Saldanha Bay, the last port of call in South Africa, before the strong offshore winds set in. In hindsight, we should have stopped at the Cape. I had assumed that, in an offshore wind, if the worst came to the worst while approaching Saldanha, we could have lowered both sails and drifted offshore. Not so. At 2am, we were fighting for our lives, close hauled on the starboard tack trying to get into the yacht club anchorage. In the darkness, and the screaming wind and rain, I hand steered as best I could to keep off the rocks as the swells behind us started to break; there was no longer an exit. This was one of the rare occasions that we wore our lifejackets and clipped on to avoid separation had we finished up in the rockery! Ahead, I could make out the flashing red and green channel markers, but couldn't gauge how far away they were until they passed alongside way above my head. Nicole was locked inside plotting our course on the chart and shouting directions, to be interrupted by the pilot boat, warning that a giant ore carrier was about leave the terminal ahead. Nicole replied, "We are hampered in our manoeuvres, and will maintain our course and speed." This was accepted by the pilots. It was with great relief that we picked out a vacant mooring with our searchlight, made fast, readied the kedge anchor just in case, lashed up and stowed. I threw back a warm (no fridge) gin and tonic, had a hug and we crashed into our separate bunks. The club was a brilliant DIY place with its own slipway for hauling out, plus helpful and knowledgeable members. They couldn't believe that we had made it into the bay in the conditions the previous night. 'Beware an onshore breeze puts you downwind of the pilchard plant'; what a stench. We loaded up with fresh fruit and veg and extra water, ready for a long passage across the South Atlantic to Brazil, and took a taxi to the ore terminal for a surprisingly convenient one-stop check-out with Customs.

Nicole's desperate chocolate cake

No eggs? No butter? No oven? No problem!

2 cups all-purpose flour
1 cup white sugar
1/3 cup cocoa powder
1 teaspoon salt
1 teaspoon baking soda
1 teaspoon baking powder
1/3 cup vegetable oil (anything but olive oil)
1 tablespoon vinegar
1 teaspoon vanilla flavouring
1 cup warm water

Line a heavy duty frying with baking paper. Mix all dry ingredients in a large bowl. Make 3 wells using the back of a spoon. Pour oil into wells, followed by vinegar, vanilla extract and warm water. Mix carefully with a spoon. Pour mixture into the frying pan and cover with a snug-fitting lid. Cook on stove top at the lowest heat possible for around 40 minutes. Move the pan frequently to prevent burning. Once you see little holes on the surface (like crumpets or pancakes) and the cake springs back when lightly touched, it is done. But do not lift the lid until at least 35 minutes have elapsed. I use a glass pan lid so that I can see what is happening.

Cool and turn over. For easy icing, use icing sugar, water and vanilla extract.

If you happened to have 3 tablespoons of butter, you could make icing adding 2 cups of icing sugar and 2 tablespoons of cocoa to it.

The longest trip

As we cleared the entrance to Saldana Bay, the autopilot went haywire. "Shall we go back to fix it?"

"Nah, even if we have to hand steer all the way across, I don't want to go back; it's taken us a long time to get around Africa," Nicole replied. We settled into the best and longest passage of all times. I figured out that it was the electromagnetic compass that was malfunctioning, and swapped it for an old spare to get Mike, the autopilot, steering for six blissful weeks of gentle, warm reaching across the ocean. The best route was a sweep towards the equator passing close to St Helena before heading back down to Salvador da Bahia. We did not want to spoil the magic by stopping at St Helena as most yachts do. We had everything that we needed aboard plus the occasional top-up of fresh water from squalls in the night. Nicole was a bit freaky about water supplies and could sniff a squall in her sleep. Sleepy-eyed and stark naked, the water ghost would appear on deck in the freezing rain and carefully arrange bits of string to direct the flow of water from the awning into all the buckets and bowls that she could find.

Night watch after a squall

By midday, the water in the shower bottles was hot enough for a shampoo. The bottles were one-litre plastic bottles with eight fine holes drilled in the lid; efficient and economical. Occasionally, we would collect enough water for a bubble bath, improvised with a plastic sheet laid over the forward trampoline. Before reaching Salvador da Bahia, we crossed Taraipo's outward track and I had completed my circumnavigation, which left me melancholic for the night watch, alone, as I could not think why I should continue voyaging. I had done 44,000 miles in four and a half years and only spent one night ashore. Nicole came up with a reason the next day. "Let's sail to Opua so that I can put circumnavigator on my CV!" *Aha,* I thought. *She is still keen to continue.* For the record, it had taken us five weeks and six days to reach land again.

Nicole – communication

I was very happy that we did not have e-mail or a satellite telephone. We had a SSB receiver only – so the only possible communication we had with other boats were the VHF radio if they were a within 30-mile range. We knew what was going on in the world, were able to listen in daily to the 'Rag of the Air' and other forums and download weather maps. But we did not get any personal messages and could not write to anybody. What freedom! A freedom that a lot of people give up for the sake of, in my opinion, false security. We had an EPIRB in case of an emergency, to be triggered once, and then rescue would come as soon as possible. John and I often discussed 'what ifs'. Like, what if one had a heart attack while at sea? Well, that would mean that, hopefully, the casualty would get transported to a doctor somehow and the other would have to continue sailing. Even more important to have a boat that I felt confident to sail on my own.

Latins

In the marina, we moored next to a cheeky French single-hander who invited us for crepes the following evening – after a busy day shopping for a new bikini and a few other essentials. Christophe, late fifties, specs with a bit of a leery look, was keen to see the new acquisition. Nicole obliged, and our host brought out his camera while waiting for the crepe mixture to settle. It turned out that he was not single-handing by choice. His transatlantic crew had left him in Rio to sell bread and he was keen to find a new girl to fill the vacant forepeak. The next day, he presented Nicole with some flattering prints from the photo shoot of the bikini. It was some time later that she realised that he had Photoshopped her boobs to great effect! We loved the Brazilians, but you had to watch your back, especially in touristy areas. We never wore jewellery. I carried only one debit card and just enough cash for the day, zipped in my pocket, and purchases in a backpack, which could not be snatched. In a crowd I would carry it on my chest. Chillingly our flashy neighbours were attacked between the marina and a nearby restaurant. Another sailor was distracted by a clown while his accomplice came from behind and with lithe hands emptied both trouser pockets, he did not notice that his wallet, and cash had gone till he got to the next shop! The French crews said that Itaparica was an island we had to visit. We worked our way down the channel between the island and the mainland to a secluded anchorage off a large hangar ashore, the home of catamaran and seaplane builder Robert Dotta. In the middle of nowhere we were amazed to walk into a hangar full of hi-tech moulds and machinery for building the latest 'A Class' racing catamarans, cruising catamarans and sea planes, all to Robert's self-taught designs. At low tide, he helped us with some stainless steel welding on Taraipo's worn rudder fittings. At Sunday lunchtimes, Robert's local clients would drop in between us and the shore in their seaplanes.

We felt that we could have stayed there forever. In the nearby village, there were waterside building plots for sale, some of which had been bought by German sailors. However, we became aware of an uneasy atmosphere in the village as we went to buy our daily bread. There was only one bus per day into the city, which left at 6.30am with the school kids, and did not return! Everyone, young and old, had to walk back through the rain forest, the two miles from the main road. Too scary for us to contemplate in the dark, and not even in daylight after the beer salesman who came once a week to collect the order from the bar was killed for his moped. Everyone in the village knew who had done it, but nobody was telling. A few days later, the culprit was killed in a shoot-out with the police; a cost-effective justice system! Another disconcerting thing about the anchorage were the vibrations through the hull when the locals were out fishing with explosives at night; a totally destructive and banned practice, and everybody knew which young guys were doing it, but nobody told, and the fish were sold.

Our immediate concern on arrival in Natal was to find a stainless steel welder, a big stainless bolt and some rubber 'pit belting' shock absorber to refasten the main crossbeam to the starboard hull. The bolt had snapped at sea. Broken stainless steel fastenings and fittings were a frequent problem on the voyage, and no longer came as a surprise to us. Fortunately, Pieter, the builder, had used the highest-quality materials, and James Wharram, the designer, had incorporated enough redundancy into the structure to enable us to make harbour safely each time. Nevertheless, it was very scary when these things happened and there was no guarantee that we would always have the ingenuity to make it back to land. We had fun chasing up our unusual requirements with the local garages, who were generous and helpful. For the last evening ashore, we decided to spend our remaining *reals* in a *pasteleria.* We showed the owner our last reals and he brought us poor sailors a *caipirina*, a jug full of freshly squeezed watermelon and more pastries than we could eat.

Nicole – cruising community

It was the first time in my life I felt part of a community – the cruising community. The most interesting people are sailing, are 'live-aboard', families, couples, singles, nuclear engineers, bakers, psychologists, firemen, electricians, you name it. Of course, you never like everybody, but you all have something in common: at one stage in life, we had to take the step in the unknown, outside our comfort zone, and we realised a dream – to go sailing. We all had to earn our good time at anchorage with exposing ourselves to the sea, to weather, and mostly to ourselves. If you met someone you liked, you invited them to drinks or dinner and most of the time you had a great evening telling stories and helping each other with knowledge and manpower if necessary. Sometimes we met the same people two years later without any contact in between and it was like meeting old friends. I simply loved the life on a boat. I did not always love sailing, but I always loved arriving, and there is nothing like going at three knots downwind, flat seas, and having the horizon all around you.

We celebrated Bastille Day, my birthday, in Degrade des Cannes, an unlikely suburb of Cayenne in French Guiana, by hitching a lift into town and buying a baguette, a camembert and a bottle or two of Bordeaux wine. We hitched a lift back to sad Degrade and shared our spoils with the only other visitors in the river, Regis and Babette aboard Xe from Noirmoutier, whom we had met in Natal. They had lost the top of their mast when bumped by a ship while they dozed on watch! We visited the famous prison island 10 miles offshore and saw the pond where prisoners were thrown to the crocs. We only stayed one very bouncy night in the anchorage, where we were noisily attacked in the night by a stream of coconuts knocking against the hull as they drifted by when the tide reversed.

As we headed for Paramaribo, we saw an unusual ship drifting and waiting to retrieve European space rocket capsules that had been fired from Kourou. Later, we went to the rescue of a fisherman adrift in the middle of the ocean who was waving his shirt on a stick. I decided that we could not help him; there was a heavy sea running and he was drifting towards the distant shore. I put out a 'Pan-Pan' message for him, but felt some guilt as we disappeared over the horizon. I was confident that he would be picked up by the coastguard or fishermen in the area.

Since first hearing the name, I always fancied sailing to Paramaribo. We were faced with a mass of shallow sandbanks and swirling sandy currents ringing the coast as we struggled to make the entrance. Once inside the river, we were not disappointed with the detour to Suriname, about which we had scant information. As evening fell, we ventured into shallow waters in order to anchor out of the current. Tidy properties ashore had an unmistakable civilised Dutch feel to them, but nobody waved. The next day, we sailed upriver, pushed by the tide, and anchored with difficulty in the muddy current in the middle of the capital, but not for long. While we were having lunch, I suddenly realised that the scenery ashore was passing by rather quickly. The tide had reversed and we were dragging our anchor through the soft mud. We managed to re-anchor near the only landing possible, which led awkwardly through the grounds of a luxury hotel. We loved the old city, but decided to look for a safer anchorage upriver, which we found in the form of a permanent mooring in the ripping current off Domburg.

We needed teak to replace some of the slats in the deck, and a non-return valve for the engine water cooling intake. Amazingly, a nearby Chinese store stocked brass non-return valves. We could not believe our luck. The next mission was to persuade one of the saw mills that receive endless barge loads of Amazonian timber to cut a few very small planks for Taraipo. Nicole took her purse, walked upriver and was lucky to find a mill owner's son who was prepared to sell her a few offcuts if she went back after 5pm. He even sanded the boards for her. "Well done, Nicole." She was not quite so successful on the Sunday lunchtime when rowing ashore to collect a cold beer for the skipper at the crowded riverside bar. With her red purse on the seat, she 'caught a crab' while pulling hard against the current, fell backwards, capsized and, to my dismay, drifted down river on the fast-flowing ebb tide. Panic! There was nothing I could do to save her. Fortunately, a passing fishing boat saw her predicament and came to the rescue. They were in for a surprise, as she was only wearing a sarong when she fell in, which she did manage to hold onto, and the oars, and Tuggii the dinghy, but not the red purse. The guys told her to beware of piranhas, the river being like the Amazon. 'Spoil sports'; that was the end of cooling 'freshwater baths', hanging off the boarding ladder in the roaring current. The freshwater also removed all traces of marine fouling from below the waterline, which saved us the cost of a haul-out. Despite taking the utmost care and drying our inner ears with a tissue after every swim, Nicole caught an infection. We found a public clinic nearby, clean, organised, and she was treated within the hour. With two further check-ups, the total cost was less than $10.

I get a real buzz from the challenge of manoeuvring under sail, and take every opportunity to enjoy the right to travel the world's waterways with the engine switched off. It's quieter, cheaper and more fun, but has the potential to be embarrassing. We set off downriver with the first of the ebb. By the time we hit the busy city, we were tacking close-hauled at 10 knots, much to the surprise of the wee overloaded ferries, misjudging their avoidance tactics. You can't win them all. Within the hour, I met my 'come-uppance' snapping smart orders to Nicole at the helm as we drifted on the current attempting to return to our first anchorage near the wide entrance to the river. I tacked too late and we finished up impaled on a pole in a fish trap. This had the potential to make a hole in the bottom of Taraipo as we sank down with the tide. Fortunately, with a little adrenalin, we managed to heave ourselves clear.

Almost like home

The Customs and Immigration service in Trinidad and Tobago had British rigour with a Caribbean twist. We arrived in Man O' War Bay in the dark at 3am and gingerly tacked our way into the anchorage, guided by the Charlotteville sector light. We rowed ashore at 8am to be told by Customs that we had to pay a fine for not checking in on arrival! As the ATM nearby was not functioning, the officer lent us from his own pocket the bus fare into town in order to withdraw cash to pay the fine at head office. You can't argue, so we spent the whole day dragging into Scarborough, the capital, and back by bus. Wonderful scenery and super-friendly inquisitive locals as I travelled with 'my daughter' (good for my ego; not so good for Nicole's) through the Scottish, Welsh and English counties of Tobago.

Because our boat was a bit 'ethnic', and because Nicole was often bottom fishing off the bow, the locals cruised by closer and closer until one day Taffy, muscle bound and black, stopped alongside and, with a broad grin, invited her to go fishing with him for the day. She couldn't say no! When they returned, he respectfully asked me if was okay to cook some fish head soup. He then stripped off his T-shirt and the pair of them were trying to avoid each other in Taraipo's tiny galley. They had had a great day's fishing. I didn't suspect that there was any more to it than that. However, I was pleased to see him head ashore to his own folks. The locals all fish for kingfish single-handed, in high-speed dorys with two outrigger rods. Their technique is to cruise at nine knots through a shoal, trolling bait.

At the first stopover on our way south, Nicole decided to sell some shell jewellery to the tourists on the beach. So off she went in Tuggii with her sarong and a basket full of necklaces, bracelets and anklets for sale. Through the binoculars, I watched her spread the sarong, but she didn't seem to be attracting many clients. Then she disappeared into the bushes with her wares for a long moment, leaving Tuggii tied to a tree. Apparently, the 'Ice Cream King' was doing a roaring trade, and had offered Nicole $600 and a drive around Tobago for all of her jewellery. She turned him down! In Bucoo Bay, we heard that there was to be a 'Donkey Derby'. As we were preparing to row ashore, I saw a black guy paddling our way on an old surfboard; guess who? Taffy no less, dressed up to the nines in brand-new trainers, Adidas T-shirt and gold necklace. I could see that I was in for a difficult night at the ball, which turned out to be a big noisy party with hundreds of islanders taking part in the goat racing and liming. Taffy disappeared into the crowd. He was quickly replaced by a smooth dancer. Obviously not impressed with my moves, he put a cold beer into my hand and took the hand of my partner, leaving me stranded in a mass of heaving bottoms. "If you can't shake your bum, you ain't never going to get far, no matter how much you been drinkin' de rum." Not good for my ego – I let Nicole row us home that night.

I was back after four years' absence in Chaguaramas anchorage, Trinidad, where more than a thousand yachts per annum are serviced. We reunited with lots of cruising friends, and set about resolving many technical issues with the availability of plentiful supplies and yacht expertise ashore and among our neighbours. As we ferried back and forth in our ancient PVC inflatable dinghy, a neighbour noticed that one of us had to keep pumping it up en route in order to stay afloat! He called us on the VHF and told us he had a dinghy for sale; were we interested? I paddled over to see what the story was and, for a £100 (£1,500 new), I bought the perfect dinghy for us, a big seaworthy and virtually indestructible Hypalon Avon (PVC seams melt in the tropics). Having done the deal, I was keen to inspect his yacht, an unusual long, low-aspect rig, one-off cruiser. He was quite reticent, did not want to discuss the special features of the design, but politely showed me around and introduced me to the almost-mute crew members who didn't 'chatch like normal yachties'. The new dinghy meant no more taxi boat calls on the VHF radio – 'Chaguaramas shuttle, Chaguaramas shuttle, Chaguaramas shuttle'; a bit of a mouthful after a few drinks!

At that time, I was having big problems with the diesel engine. It was becoming almost impossible to start, requiring lots of 'Quick Start' into the air intake and sending clouds of toxic black smoke out of the exhaust.

Nicole and friends had a long-term plan to meet up in the Grenadines for the Christmas holidays, so we had to leave without the engine. Reassuringly, I famously stated, "There's always wind in the Caribbean." Our good friends John and Julia and dogs towed us out of the Bocas del Dragon with their powerful tender, to release us into a dying wind and lots of current drifting us off towards Venezuela; there was nothing that we could do to remain on course to Carriacou. The next morning, scanning the horizon in the shimmering mirror calm, 50 miles offshore, I thought I saw a tree trunk floating; through the binoculars, it turned out to be a man, waving! He could swim towards us, as we were not able to move. I could have launched the dinghy, but he would have capsized me in trying to climb aboard from the water. He was soon with us, floating on a yellow mattress; a fit young man, clean-shaven, and smartly dressed in T-shirt and shorts. He didn't want a drink or food. His story, in Spanish and tricky to decipher, was that he had been out night fishing with his dad and two other men in their 'panga' when they were attacked by bandits who wanted to steal the 200hp outboard motor. The father resisted, the bandits shot him and his crew, and 'our boy' slipped over the side. When daylight came, he found the mattress in the debris and set off for Taraipo. A very scary scenario for us. My guess was that they were drug smuggling, as this coast was the main exit for Columbian drugs, as the Americans had blockaded Colombian waters. Our man, now dry in the hot sun, was persuaded to go below for a nap in Nicole's bunk. I was worried that we had a stowaway with no papers aboard. He was a very nice young man. On two occasions, he was up for swimming ashore to save us any trouble. When he surfaced from his nap, he was shocked and surprised to find us sailing 'the wrong way' away from Venezuela under spinnaker. He could not believe that our boat could move without an engine; also, it would have been mission impossible for him to get back home to Isla Margarita via Grenada and Trinidad without papers or money. He was desperate to tell his mum on Islas Margarita what had happened to his dad. We tried calling the coastguard to no avail. The bandits were also from the boy's home port, Margarita. "The plot thickens." Our man was going to be spending the rest of his life avoiding being killed by the bandits from his own town. We tried to offload him onto a Venezuelan trawler, but they refused to take him from a 'raft with a red ensign'. The wind then did a 'one-eighty' and we headed for Los Testigos, north of Carupano. As we approached the off-lying rocks, the boy again volunteered to swim the five miles ashore. He knew that there was potential for trouble ahead for us all. We drifted through the islands flying the spinnaker, the red ensign and the yellow letter 'Q' signal flag 'requesting free pratique', past a Customs gunboat dressed overall for the presidential elections. There was every chance an officer could have noticed us before we were able to offload our man into a friendly panga from Margarita. Imagine if we had been caught with a drug runner aboard; I would certainly have spent some time in a Venezuelan prison cell. Phew! We had escaped, but not for long. As we exited the islands, night fell, the sea breeze kicked in and we were sucked back through the rockery towards the mainland shore and out of the west-flowing current, much to the approval of Nicole, who was also ready to jump ship when we hit the mainland in order to be on time for the Christmas party. I was saddened to think that she was considering leaving me yet again – for a Christmas party on the island of Carriacou, Grenada. Travelling there would have been a tall order; she would have had to hitch through bandit country, across the Bocas to Trinidad, then fly to Grenada and take a ferry to Carriacou. As the atmosphere aboard was getting a little tense, we decided to try to start the engine one more time. In a cloud of ether, it fired. I shoved it into gear and kept chugging for the next 36 hours. We were lucky that, on arrival in the dark, another yacht had heard us calling the harbour master on the VHF, noticed that our navigation lights were heading for the reef and warned us to turn to port! The GPS waypoint didn't quite match the old paper chart.

After three weeks partying ashore, having time for each other and living the most opulent lifestyle, we found ourselves back aboard in Chaguaramas getting ready for the arrival of McLaren. A passer-by on the way back from buying his Sunday newspaper came alongside and asked us what the problem was with the engine. Would we like a hand? It sounded like a godsend to me. Unlike a mechanic, he did not want to hear me attempt to start the engine. He installed himself in one of our comfortable cockpit seats and asked, "Who has done what to the engine in the last twelve months?" He made notes. "Now try starting." We removed the engine hatch cover and Nicole squirted WD40 into the front cylinder air intake while I turned the key. It did not start. "Okay, you need more power to spin this baby faster. I have to get back for lunch; I'll come back with a plan this afternoon." Off he went to his classic 50' Hatteras motor yacht. True to his word, he came back with two solutions. "You need to double the spin of the starter motor by doubling the voltage from 12 to 24 volts. The options are one, cheap and complicated using clothes pegs and rubber bands, or two, simple and expensive purchasing a diodes and a relay."

"You have to reconfigure the wiring temporarily to start, then switch back to 12 volts DC in order not to fry all the electronic instruments, radios and navigation lights." I can never understand electricity at the best of times. Nicole and Ron drew a nice wiring diagram and I organised the clothes pegs.

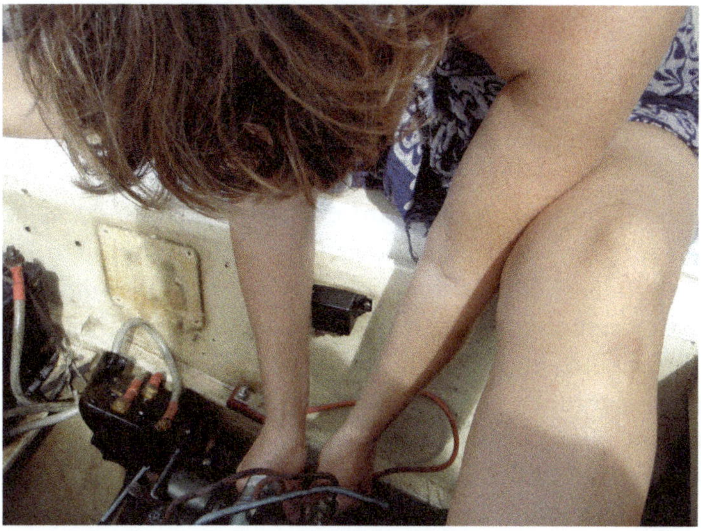

Restart wiring

The engine had to start and did so, blowing lots of rubbish out of the exhaust pipe. It turned out that Ron was in the motor trade, and insisted on his teacher wife having an extra battery in her car boot during the icy winters of the cold north-west US to be sure to start after school.

While awaiting McLaren's arrival, I received a call from Immigration at the airport saying that he had arrived and they would be sending him back to Curacao, as he did not have a return ticket. Our plan was that he would be sailing to Curacao with us. In a panic, I dashed to the airport, signed an immigration form and, finally, McLaren was allowed to enter Trinidad.

We sailed away in style, passing through the anchorage in Scotland Bay, with McLaren on the bow playing *Flower of Scotland* on the bagpipes. Next thing, we were flying out of the Bocas at 11 knots, close-hauled, just scraping past the rocks to port. McLaren was below carefully packing his pipes while Nicole white-knuckled it at the helm.

Porlamar is a well-protected and spacious anchorage in Islas Margarita, overshadowed by the shell of a modern hotel with an armed guard at the door. We were protected from the prevailing wind, not from the marauding locals, who operate at night stealing dinghies and outboard engines. The only safe place to land is at Juan's Marina, a single jetty with a watchman, which accommodates 30 or more dinghies by day. Juan and his mum had a monopoly on the yachting trade, doing all the laborious Customs and Immigration paperwork, organising minibuses and chaperoning supermarket shopping trips. We shopped downtown, which was more fun, much cheaper, cash only with black market Bolivars exchanged for US dollars from furtive guys to be found in the shade of every street corner. "Keep your hands in your pockets, McLaren." He couldn't resist buying a bikini for my girlfriend; that's mates for you! I got a pair of flip-flops. We loaded up with cheap diesel, cheap gas, cheap beer, wonderful fresh fruit and fresh veg, and headed for the deserted anchorages of Los Roques archipelago via Isla Tortuga to collect shells on the windward shore for jewellery making.

Shells for jewellery making (photo: McLaren)

Hand steering into Los Roques

Nobody lives on La Herrera. It is one of the rare anchorages that is well protected and has a clean wind for windsurfing. You can windsurf forever in Los Roques and find deserted anchorages. The only disturbance we had was a group of kite surfers who had surfed all the way down from Margarita accompanied by a 'dormitory' yacht; a nice idea.

McLaren lands one

Between Aruba and Curacao, we were buzzed by a Dutch Navy helicopter with armed soldiers hanging out of the open door; they called us to identify ourselves on VHF. "This is British yacht Taraipo."

"Why are you flying the Venezuelan flag?" Oops.

"McLaren, take down the Venezuelan courtesy flag and replace it with the French one and attach it with the stripes horizontal!"

"Nicole, break out the red ensign quickly."

Oh dear, we failed our flag etiquette test; 'someone' had forgotten to take down the Panamanian flag in the night after fluffing our planned entry into Los Aves island 'to see the blue-footed boobies'. We managed to unnerve McLaren completely as we charged towards the cliffs and the blind entrance into Spaanse Water, Curaçao. We had wind and waves behind us, I was hand steering, and Nicole, with GPS and plotting on a paper chart, was calling the shots from inside the cabin. Finally, the entrance opened up and McLaren stopped asking from his lookout on the foredeck if we were 'sure of our position'.

The sun was 'dipping the yardarm' as we supped warm rum and Coke in the super-safe anchorage off Sarafundy's, the busy 'yachties' bar. While settling to dinner on deck, wearing party hats from McLaren's Christmas crackers, our idyll was interrupted by a blinding searchlight. A navy SWOT team 'armed to the teeth' was alongside Taraipo; they shoved the light right into my face. "Where are you from? What was your last port of call?"

"Err, Los Tortugas."

"Show me your passports."

"Captain, get the ship's papers." They suspected that we were drug runners. We were mightily relieved when the light switched off and the naval officer politely thanked us for our time; the interrogation was over. We were to take a bus and check in with Customs the next morning. Apparently, Curaçao was an artery of the drug smuggling trade and several Dutch warships were there to help the crackdown on the drug runners. This was all very surprising on a highly civilised island with great public transport, supermarkets and chandleries. There were lots of yachts stocking up with imported fresh fruit and veg from the decks of Venezuelan fishing boats at the town quay, and with good fresh water to be bought at the tap on the jetty at Sarafundy's. I never got used to the idea of paying for water, having been brought up on soft Manchester water from the Lake District, which apparently ran freely from the tap.

Just us

There were sad farewells as McLaren flew off, and we checked out for a rough passage to Colombia. We managed to avoid the roughest bit of ocean by sailing close inshore, much to the annoyance of the Venezuelan coastguard lookout on Islas Los Montjes del Este. The radio crackled. "Velero. Velero…" I didn't reply. They had seen us sail into and back out of their harbour, a rocky inlet, having decided that lying in the swell alongside a stone jetty would be too rough for Taraipo. We did not like the desolation of the Cabo de Vela anchorage either, and finally found peace alone in Gairaca inlet in the Parque Tayrona.

We were disturbed several days later by the arrival of a big, clunking diesel-engined fishing boat, Twilight, with owners Willie and Mary; lovely characters whom we had last seen in Curaçao. They later crewed for us through the Panama Canal. Their yacht was a converted Louisiana shrimp fishing boat. He was a retired crane driver and she a firefighter from New York. Willie loved diesel engines. They had previously owned a beautiful sailing yacht, but he would never turn off the engine, so she told him to buy a motorboat for their twilight years, which he was still converting with welding and cutting gear! They were good fun, organising regular 'Mexican Train' sessions in Curaçao, and generously gave us the freshly made water we needed to extend our stay in Gairaca.

Twilight left. We started to feel lonely and felt the urge to move on.

Rodadero was a bustling tourist resort. We had returned to the real world; it was too busy for us. We bought some stores, paddled back to Taraipo and left the following morning. Columbia did not feel like a safe country. We had to watch our backs all of the time, but it was exciting, no more so than in the approaches to Cartagena. We had decided that we would take a shortcut through a narrow gap in a submerged Roman defence wall between Bocagrande and Isla da Tierra Bomba; it was dusk and we could not make out the flashing green and red lights that should have marked the edges of the wall, against the backdrop of the city lights. Change of plan. It took us seven hours to sail around La Bomba into the narrow main channel, which is only two ships wide. I decided to sail just outside the channel markers, leading into the wide expanses of the main harbour. There were fixed (not flashing or occulting) lights everywhere. Too late, I realised that a high white light was getting closer and then a red and green; a ship dead ahead! We turned hard a-starboard, but too late. She gave us seven loud blasts, as she had had to take avoiding action (slowed down) in order to avoid a collision. My limp excuse was that I had spoken to the pilot station near the entrance and told them of my intention to sail into the Marina Club Nautico anchorage. I was annoyed; that was the second time in my sailing life that I had been hooted at. Cartagena is a stunning city to arrive in under sail and explore on foot.

We had an inglorious two-day sailing epic, being blown backwards while trying to enter Bocas del Toro, Panama, against wind and current under sail only. We had arranged to spend Christmas in Bocas, so we had to be able to leave the boat in a marina berth. None of the guide books told us about the sand flies in this otherwise-idyllic tourist location; they attack at 5pm every afternoon, by which time we had to be behind mosquito netting. It is hard to understand why so much American money is being invested in ritzy houses on stunning, but steep island plots, on land they can never own by law. We joined good friends Conny and Beat for a 'week behind bars' (fly screens) in a luxury resort for a week, and from there we moved by high-speed taxi boat to Al Natural Resort for a few days, where the bugs were even bigger and more vicious in what is a stunning holiday location. I had assumed that at 'Al Natural', we would all be taking off our clothes maybe! No, it was not like that. In fact, the four of us, good friends, were not at ease snoring together in a tropical open-plan, open-to-the-sea beachfront cabana on stilts. We were just about talking to each other on the taxi boat ride back to Bocas. There, we separated, still the best of friends. Conny and Beat headed home to Switzerland, and we stored up for the next trip.

Stowing the tool locker

A little bird told me that there was an engineer in Colon who could fix us up with a new engine cheap, so off we went for a windy sail tacking and gybing amongst the shipping of the world anchored outside the breakwater awaiting their turn to transit the Panama Canal. We listened to the control tower telling ships when to shorten up their anchor chains and when to start moving, giving us a sense as to where we could slot into the procession of leviathans and head to the yacht anchorage on the Cristobal Flats.

We met an engineer over a cold beer in the decaying Panama Yacht Club, noisy fans spinning slowly overhead. He reckoned that he could import and fit a new Yanmar diesel for $6,000, cheap. Then followed a series of setbacks and price hikes as it became clear that the yacht club would not allow the job to be done in their marina, and import costs multiplied. We were determined to transit the canal with or without engine, and we put out regular calls on the yacht net appealing to the 30 or so yachts awaiting transit to consider giving us a tow. The cost of hiring a tug would have been prohibitive and would probably have torn the bow off Taraipo, and the canal pilots were not keen to allow a yacht to tow us through. I phoned the local owner of a luxury motor yacht whose family had bought some of Nicole's jewellery. He put me in touch with a young canal pilot who suggested building a wooden frame on the stern of Taraipo and to buy a second-hand outboard engine to push us through. Then André and Babette, with whom we had spent a day in Brazil, dropped by for a coffee and offered to tow us through if we could get permission from the Canal Authority. I raced over to the control tower, got through security and in to see the chief pilot, who was very friendly and gave his permission with the proviso that we could complete the paperwork and deposit a double buffer at the Canal Company's bank that afternoon. The transit charge was $600. The $1,000 buffer is refunded if 'a speed of eight knots can be maintained' and there are no breakdowns in the locks. I had my credit card swiped at the bank for two buffers to cover any nuisance we may cause from the towing operation. Problem: I did not have enough credit on my account and the bank was closing within the hour. In a panic, we had to find four ATMs to withdraw $500 at each, stuff our pockets with greenbacks and run the gauntlet of gun-toting shop guards back to the bank, hoping not to be robbed on the way! We managed to round up four friends to be the mandatory line handlers for the 50-mile overnight trip through Gatun Lake, and bought enough food, beer and ice to feed them and the pilot. When the pilot came aboard, everything had to happen quickly; he was receiving his orders by radio from 'mission control'. We had to fall in behind Atsani and the towline became frighteningly taught as we careered downwind. We followed a container ship into the cavernous bottom lock, ablaze with light and with two other yachts attempting to 'nest'. Hats were essential, as weighted heaving lines are hurled down from above as the yachts spin in the swirling currents created by the ship and tug ahead. The pilot was cool and competent. It was not the first time he had seen yachtsmen at work, bumping around the locks. We were nested alongside a big yacht commanded by a bossy lady giving orders to all and sundry, but herself not able to take a turn on a winch when the strain came on, nearly sandwiching us against the black slimy wall as the gates behind closed and the fast-rising waters flooded in. Two hours later, we cleared the top lock into the eerie, dark freshwater Gatun Lake. The pilot went ashore for the night. We were tempted to go for a skinny dip, apart from the crocs in the water and the monkeys in the trees. At dawn, the pilot returned in a rush. He was pushing us to cross the lake and down the locks into the Pacific before nightfall. What a feeling of freedom to hit tidal waters once again. It was a big moment for us as we passed under the Bridge of the Americas, which carries the road from Alaska to Panama, but, surprisingly, no further south.

Panama transit crew and extra fenders (photo: Babette Rihoué)

Sadly, the authorities had removed the slipway at the Flamenco anchorage, obliging all the yachts to pay to land their dinghies at the ferry landing. We heard on the grapevine that the Canal Company had also arrived at the yacht club back in Colon with a bulldozer early one morning and pulled it down! Not that we cared; we were not stuck there trying to install an engine. We were being wined and dined in Panama City by a group of American expats, a tired young canal pilot and a gold trader. The trader met us at the dock and chauffeured us around the city's best restaurants, and the pilot took us for picnics by speedboat. How the other half live, insulated from the heat and the poverty on the other side of the fence. Our new friends admired our adventurous lifestyle, but could not imagine a life here in the tropics without air conditioning, a freezer, a fridge, a TV and the potential to earn lots of dollars. Using the tides, we imagined that we were going to be able to beach and antifoul the hulls on the last beach in the Isla del Rey before setting off across the Pacific. No chance. Despite there being no wind, a low southerly swell was breaking on the shore. In attempting to go beach combing, Nicole had a scary ride through the breakers in Tuggii and had to walk him, like a dog, the length of the beach in order to find smaller waves to get back out, much to the amusement of sailors aboard other yachts preparing to leave for the Galapagos. I had a grumbling groin hernia when we arrived in the Isla da San Cristobal and was referred to the nearby Ecuador Navy Hospital for a check-up. The surgeon said that she would not operate at this stage, and suggested that I should be able to sail to the Gambier Islands before surgery would be necessary. I paid my 12 bucks and, smiling, wandered back along the seafront dodging the enormous fat, smelly sea lions that have invaded the promenade. We had not planned to stay for long in the Galapagos Islands. The authorities had put in place strict controls on yachts to prevent the desecration of these extraordinary islands by the tourists that we were. No blue-footed boobies, but we did get our fill of water, and the best un-chilled (they last better) vegetables ever for the long haul to French Polynesia. It took a day and a night to clear all dangers, until a minke whale passed us and then returned for a closer look. He must have liked what he saw. He called a mate and they proceeded to swim and blow bubbles under the boat! We were petrified for an hour as we held on and took photographs, and then they glided away quietly. Two days later, there was a large tuna fishing vessel parked on our track. Nicole called to inform him of our intentions; he came back and surprisingly offered a fish. I could not imagine how he imagined that we could go alongside to collect it in the rolling swell until I noticed a yellow speedboat being lowered over the side with two men aboard. In no time, they were bumping our fragile topsides with their roughly driven aluminium speedboat, and throwing a massive kingfish aboard. There was also a plastic bag containing a tourist guide to the Ecuadorian port of Manta plus an e-mail address for our radio operator. Lonely captains never miss a catch.

Back in the Pacific

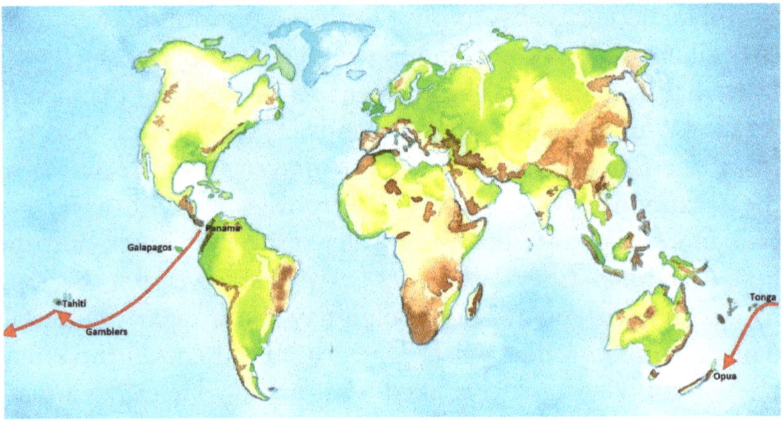

The Gambier Islands were not as I had imagined. At 20 degrees south, they are colder and wetter than the Marquesas. The southerly approach unusually takes you eerily over the top of the barrier reef, not through a pass, and then over an inner reef to an enormous sheltered lagoon full of black pearl farms, then follows a zigzag passage through the coral into Rikitea town on Mangareva Island. I stupidly had assumed, without checking, that there was a hospital in 'the capital'. When the nuclear testing range on Muraroa (500 miles away) closed down, the French Navy left town and took the medical facilities with them. They also took the heart out of the populace, leaving them no longer willing or able to farm and fish. They are now totally dependent on the fortnightly ship bringing French produce from Papeete a thousand miles away and subsidised, aluminium prefabricated houses, made in France.

Les Gambier – happy to arrive after crossing the Pacific
(photo J. Paillardon)

Islanders requiring hospital treatment are flown to Papeete. I decided that my hernia could afford to wait. We stayed longer than planned in the Gambier, as the wind was blowing against us from Tahiti. The locals were friendly, we both speak the language and there were stunning sheltered anchorages to explore. We located the only vegetable farmer still in business, so we had plentiful supplies of fresh fruit and veg if we were patient – we had to walk up the hill behind town, find the crops and wait around for the farmer to turn up. His other job was to deliver sandwiches for the daily plane to Papeete.

Soon, we were exploring the pearl farms and workshops with a view to adding to Nicole's shell jewellery repertoire. The pearl farms depend on Chinese women being flown in annually to graft a pearl into each mussel, fastened to long ropes that are suspended from hundreds of buoys in the deep lagoon. The local men have the back-breaking task of hauling up the ropes to clean the pearls regularly, and diving to harvest them.

Searching for shells in the Gambier Islands

We were invited for a day sail and picnic with Hans and Heike on his massive 21-metre-long catamaran. Hans is one of the toughest guys I ever met, steely, strong and smart. I discovered him the first time quietly anchored under the cliff in Opua away from all the other visiting yachts, though he would never classify his boat as a yacht or himself as a yachtsman. I would call him a buccaneer. He traded under sail with his father in the Indian Ocean as a youngster, and was inspired to build this incredibly physical catamaran with native boat builders on the beach in Gambia. The rig that he chose was the Polynesian crab claw; only a very strong single-hander could handle it. As well as all his traditional skills, he was one of the first sailors to use Google Maps to find his way into hidden anchorages, while voyaging and trading his way around the world.

Nicole was the first to swim towards the shore, to be followed by a bevy of small sharks. I called her back. With several youngsters in the water, I did not want to test my theory that sharks don't like to eat human flesh. We could have lingered longer with so many new friends and great anchorages to share, but needs must, and the hernia was needing attention.

When the south-east trade wind returned, we plotted our way through the Tuamotu atolls, most of which were invisible, as they are so low in water. Eleven days of gentle sailing under spinnaker sun all day, moon and stars all night towards the hospital in Papeete.

The Immigration officer in Papeete told Nicole that she could not extend her stay in French Polynesia with a Swiss passport unless she made an application from abroad and paid a hefty deposit, refundable on departure!

Fortunately, the young Parisian surgeon operating on my hernia was very understanding and wrote a note to Immigration explaining that I would require 24-hour nursing cover. Problem solved. We put down both anchors on a pristine coral seabed, ready for my convalescence. Inevitably, after several weeks holding firm, a gale blew up and we were blown off the steep coral bank into deep water. Having let go all of the chain and rope that we had, we eventually dragged to a halt in the middle of the main channel and some boats on permanent moorings. At this point, my heroic girlfriend, in shorts and bikini top, dived over the side rope in hand to tie us to a vacant mooring buoy. We finished breakfast, took a deep breath and set about retrieving 200 metres of chain and warp, and two 15kg anchors, by hand; we didn't have a windlass (too heavy and too much battery power required). Since that time, we replaced the Delta anchor with a Rocna anchor that has much better holding power.

We decided that was the moment to move on. The nurse had been out to remove my stitches and advised me to take it easy. A short cruise through Polynesia with Nicole doing the hauling and heaving should not have been a problem – and it wasn't.

Moorea from the convalescent's anchorage

Bora Bora; vertical clouds

We stopped in Raietea for an expensive haul-out. The poisonous antifouling paint that we use below the waterline to stop marine growth attaching to the hull and slowing us down erodes away in time and requires repainting every two years. In tidal waters, we used to beach the boat at high tide and wait for the tide to drop before scraping and painting the hull, usually on more than one tide.

In the tropics, there is virtually no rise and fall; therefore, an expensive haul-out is necessary. We found a low-tech boatyard (a tractor and flatbed trailer, and two big Polynesian swimmers to guide Taraipo onto it) on Raiatea atoll. Once out of the water, we set to pressure washing the remaining old antifouling paint before sanding and repainting both hulls (42 metres of waterline in total), all completed in four days with no shade from the burning sun.

Tsunami

We couldn't sail past majestic Bora Bora again without stopping. During our passage across the lagoon, and into water shallow enough (10m max) for anchoring, the sky covered and it started to rain. As dusk fell, there was a tap on the hull. "We have visitors." The owner and skipper of a Scottish superyacht that we had 'saluted saltires' with previously had come armed with a bottle of Scotch. We chatted solemnly in the warm, pouring rain, 'downed the malt', and were then transported to their yacht for showers and dinner prepared by the cook/mate. How generous. They were interested to see our unusual yacht, and amusing to yarn with. That night, we received a tsunami warning with a recommendation that all vessels should leave the lagoon and head for the open sea! At dawn, pretty well every yacht and ship left. I reasoned that we were safely anchored in a narrow, twisting channel with about half a mile of flat-topped coral reef, la Recife du Motu, between Taraipo and the ocean, which would absorb the mightiest wave. Nevertheless, we did lay all four anchors and the kilometres of warp that Taraipo carried, ready for the ultimate storm! In fact, the wave was much smaller than predicted and we didn't feel it at all.

We tried to land at Palmerston Atoll to give them a big mahi-mahi that Nicole had caught and was far too big for us to eat. Sadly, the wind was blowing into the anchorage, which would not have been safe. We pressed on into a forbidding sky and a strong south-easterly wind to Neiafu in the Va'vau Group of Tonga, one of our favourite cruising grounds, where there is plenty of wind and flat water; a sailor's dream.

Completing the circuit

"We are nearly there." Near to completing Nicole's circumnavigation; only 1,000 miles to New Zealand and for sure a gale to be negotiated. As it turned out, we averaged 70 to 80 miles per day and finished in the Bay of Islands in fine form. We dressed Taraipo overall and had a great party in the rain; we drank champagne and dispensed John Jameson Irish coffees to keep the 30 circumnavigators warm whom we had crossed tacks with on our way around the world. Sailors will note that the deep V hull sections of the Polynesian catamarans are great load carriers. One fellow cruiser who missed out on the party was our French friend Christophe, last seen in Salvador di Bahia, Brazil. I overheard on the radio about a yacht, La Belle Mumu, calling the NZ coastguard with a 'Pan-Pan' message explaining that he had dismasted 40 miles out and that he did not have enough fuel to make it into Opua. The coastguard organised a rendezvous with a ship that was, surprisingly, able to transfer 100 litres of diesel to La Belle Mumu, and a RIB to chaperone them into the Bay of Islands. We hired a car and I drove Christophe (he couldn't face driving on the left like a Kiwi) to the riggers in Whangerai, where I had noticed a mast hanging in the rafters, which turned out to be a perfect fit. It was second hand and came with a boom, and we got it for a good price. Christophe then had to motor La Belle Mumu from Opua to Whangerai to have the mast stepped. I was very pleased to have been able to help a friend in need.

Circumnavigators' party in Opua

Taraipo was not allowed to haul-out in Ashby's Boatyard or enter the Opua Marina without insurance. I had not had insurance since the Cape Verde islands, when about to cross the Atlantic; my insurer informed me by e-mail that the policy covered daytime sailing only! I then discovered that no insurer would give me cover, because my boat was too low in value. However, there was an insurer in Opua who was able to give me temporary cover to haul-out. I had no insurance for the totality of my circumnavigation, which does make for a prudent sailor.

The engine dilemma was finally resolved when we bought a 20-year-old Yanmar that was gathering dust in the engineer's loft. They would not allow Nicole to change the engine colour to red, which she fancied. She was, however, privileged to be allowed to meticulously clean and paint the silver Yanmar in their workshop. That's where she was given the nickname Energiser Bunny. Having an engine that could drive the boat forward, across a reef-strewn lagoon in a gale, or negotiating an unmarked reef entrance in strong current, subsequently proved to be a tremendous asset. We always tried to be 'ready to sail' if the engine died.

We split that winter. Nicole had some baby minding to do for friends back home in Switzerland, and Taraipo was needing some 'structural reform'. With the boatyard staff and cruiser friends, we managed to remove the damaged main crossbeam and the yard laminated a perfect Kauri replacement.

Damaged main beam

It was difficult to line up the bolt holes to in the beam and the cabin fronts, as the hulls had sagged apart. A delicate jacking operation was required, during which the boatyard guys deserted me, as the whole structure started to creak and groan! They were not convinced that the beams would take the strain!

Much more dangerous, the painters aboard the 50-foot yacht next door lit the gas stove for a brew and blew off the deck! Fortunately, nobody died. The younger of the men tottered on deck, down a ladder and straight into the sea to kill the flames. The other man was helped down, badly shocked. Both masts had to be craned off the next day before anyone was allowed back onto the disconnecting deck.

There were no buses to Whangarei, the nearest shopping centre, so a lift in a borrowed car with a cruising family I had discovered on the sailors' radio net seemed a cool option. We waited in the cool morning air until the father finally arrived with the jalopy, its grotty roof lining drooping ominously as we sombrely took our seats. At lunchtime, I went to see if any yachts of interest (friends) were in harbour; there weren't, but there was a skin clinic. As I had time to burn, I went for a check-up. As I rolled onto my tummy, there was a deep drawing of breath from behind me. "What is that?" the surgeon asked his assistant. A significant piece of my skin remains in NZ. For the return trip to Opua, we had a rendezvous at teatime in a supermarket car park. I found the car with the two young boys inside, but no parents. Next, I saw the father being shadowed by two security guys who guided him toward a police car that had just arrived on scene, blue lights flashing. Mum arrived and drove us to the police station in order to retrieve her husband, who had been charged with shoplifting. It was late before he was finally released, complete with a criminal record for having helped himself to a few loose nuts before check-out. How very unfortunate, I thought.

Part Four: The end of the line

When Taraipo was lifted by the new Travelift at Ashby's boatyard, she weighed just 3.7 tons (surprisingly light) fully victualled and ready for the 1,000-mile trip towards either Fiji or Tonga, depending on the wind direction. We were keen to go sailing again, but not sure why, or where we were going. Conditions were gentle as we sailed the 10 miles to the entrance of the Bay of Islands, where, embarrassingly, the forestay fell to the deck, leaving the jib to hold up the mast! I secured the mast with the spinnaker halliard and returned to port for a serious discussion with my best friend who had helped me swage the all-new stays. It turned out that we had used imperial rather than metric ferrules and a good squeeze on a hydraulic press at the riggers was the solution to our dilemma.

There were no fond farewells when we left for a second time. We managed to make some easting on the route toward Tonga before being smashed by a thundering outsized wave as we lay a-hull in the night. I will never lie a-hull again. Streaming a drogue astern and drifting forwards with the breakers has to be a better solution. The power of the breaking wave crumpled the deck between the cabins and smashed the port bulwark. The broken bulwark was not a problem, as it did not affect the integrity of hull. The crumpled deck was frightening, as it created gaps through which I could see the frothing wave tops rushing between the hulls as I skipped over to see Nicole, wearing a head torch and lifejacket/harness clipped to the jackstay.

Damaged bulwark, off Raoul Island

The next day, we called the meteorological station on Raoul Island. They informed us that they could not give us a weather forecast, as they only did research. Fortunately, a ship with an all-Tongan crew out of sight on the other side of the island obliged us with a reassuring forecast for the next few days. We entered Tongatapu Island through the 'back entrance', with the strong wind behind us, leaving Makaha'a Island to port before working our way into Nukualofa harbour. The wind whistled down the harbour. We laid an anchor in the middle, took a deep breath and backed towards the slippery rocks of the breakwater, Nicole rowing ashore even faster with the stern lines; epic for a crew of two. We had a big hug, drank a warm beer and settled in for a long stay to lick our wounds and soak up the laid-back Tongan lifestyle. We were in no hurry and had plenty to keep us occupied. It would take time to source materials to repair the bulwark and deck fastenings. Time is something that we had an inexhaustible supply of. We are time millionaires. There were excellent markets and fruit and vegetables were fresh and cheap, though how to cook some of them required local knowledge. Nicole could not resist buying a 'banana tree'. She had to ask the man who sold it to her to carry it to the dinghy. Lifting it on deck without damaging the bananas or capsizing the dinghy was a challenge, as was eating the bananas on the day that they ripened simultaneously. We ate banana fritters, banana sandwiches, banana cake and fried bananas 'till they were coming out of our ears', and shared bananas with our neighbours.

We loved the Tongans. We offered a cup of coffee to a diver, but, being a Mormon, he wanted warm milk. "'I mean, what were these missionaries thinking about; there is no grass, and there are no cows on any of the atolls, for Christ's sake." Anyway, we warmed some powdered milk from New Zealand, which seemed to go down well. This poor man was diving five metres all day for sea cucumbers that are exported to China via a local agent. The diver had lived in the States, but preferred the Tongan lifestyle where 'you only work if you feel like it, as there is always food on the table'. I was not disappointed to be cruising through the silky smooth waters of the Ha'apai Group of islands again; we had a mother and baby humpback whale cruising alongside, eyeballing us through the clear, clear water – 'makes you cry'. Being a catamaran, Taraipo is very shallow, only needing 60cm of water to float in, which enabled us to moor by the fire station in the centre of Pangai, the home town of Jonah Lumu, 'the first ever Rugby Union superstar'. In Pangai, I felt accepted as an equal rather than as just another tourist, and we were respected on the waterfront as sailors. There are anchorages in the Ha'apais that we explored where we were alone for days on end, perfectly protected from the strong trade winds, from the ocean swell, where beachcombing for jewellery shells was addictive, and in the open waters behind Taraipo we witnessed breaching whales. The pilotage at speed was a challenge, as GPS waypoints didn't match the charted positions of the coral reefs. On exiting northwards, I wanted to stop at Niuatoputapu, but refrained, as the island had recently been devastated by earthquake. The yachts that were there at that time were calling the authorities by satellite phone, as communications on the island had been knocked out. There was nothing that we could do to assist and, not wanting to be in the way, we sailed by.

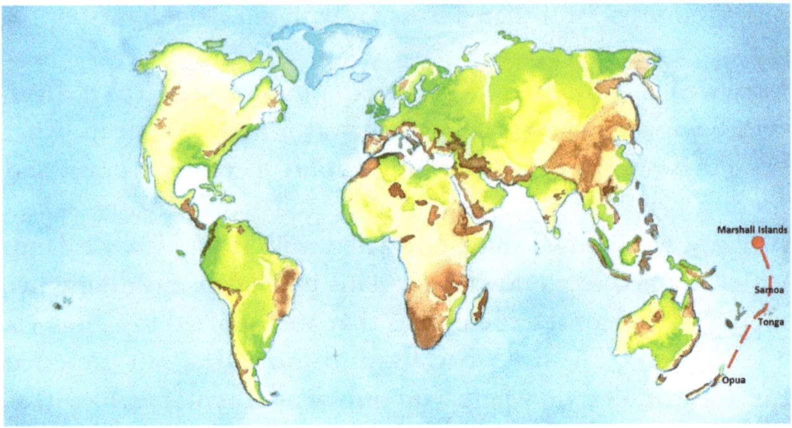

Life moves on

I was looking forward to going to Western Samoa. I had worked there briefly and had some good friends to look up. However, there are not many safe anchorages available. I thought I remembered a good place inside the shallow lagoon near to Apia Yacht Club, but the entrance was too tricky to enter. Then I discovered that all yachts have to check in with Customs in the main harbour and are obliged to stay in the marina. We were relieved to have entered in the dark following a transit of two leading lights between the breaking reefs and to be guided to the anchorage by harbour control. I went looking for my friends at the yacht club, but, sadly only found a restaurant bar and no local sailors. We did see some fearsome racing between the longest canoes in the world (100 paddlers in each), and also had a good look around one of the 25-metre-long replica Waka Polynesian catamarans built in New Zealand and donated to Tuvalu and other islands. I was disappointed not to have found any of the Samoan sailors whom I had coached, or the families I had lived with.

We had heard on the grapevine that the Marshall Islands were reasonably safe during the hurricane season, and that there are communities still trading under sail there. Good enough reasons to head that way, I thought.

Dress code

Crossing the equator, the prevailing wind and current from the east tend to push you off track westwards to the point of no return. We were lucky and had a breeze of a trip close reaching for two weeks to cover the 1,500 miles to Majuro Atoll.

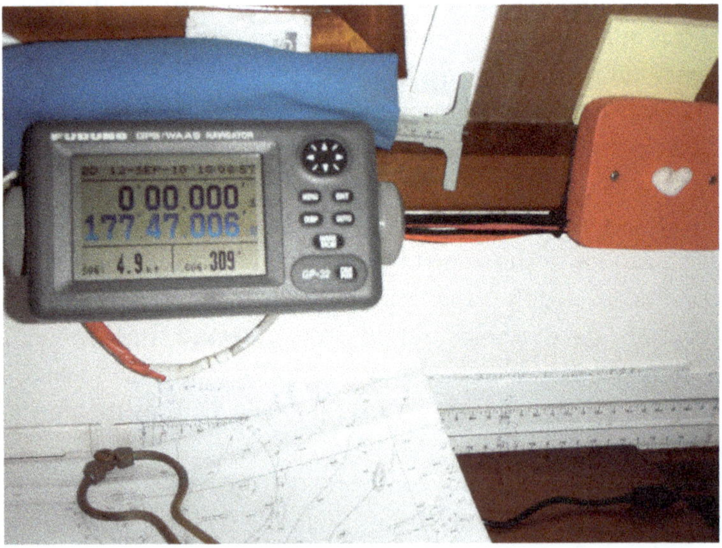

The GPS readout as Taraipo re-crosses the equator

We were welcomed by Steve, an American, ex-sideman of rock singer Bruce Springsteen, now single-handed on a steel sloop named Irony. He was the only person on deck in the early morning. He shouted directions as to which mooring to grab and, shortly afterwards, rowed over with a warm pizza in the rising wind and rain. What a wonderful welcome. We were to learn so much from him about the dos and don'ts of the intriguing Marshallese culture. When Nicole and I arrived in a new place, we did not have the urge, as most sailors do, to go ashore and visit the tourist sites. In Majuro, we did not go ashore for the obligatory entry procedures until the next day. When we did get ashore, the sun was shining and we liked the funky feel of the place. The capital is built on a narrow strip of coral about four miles long, served by a mostly Chinese (pick up until they are full) taxi service ($1 a ride 24/7). As you step ashore, Majuro has a Wild West look; the language is American, but Marshallese is the everyday language of the locals. We decided to hire a mooring for six months, giving us a secure home base to which we could return after exploring other atolls in the Radak Chain. Permission was required before visiting atolls and a certificate had to be presented to the mayor on arrival, together with a fee of $25 per visit.

Aur was the nearest atoll, an overnight sail north of Majuro; a stunning place to visit and a very tough place in which to eke out a life. Only 60 miles from the capital, Majuro, which had all mod cons, we paddled ashore to be welcomed by the children of the village who carried Tuggi to the high water mark, where I carefully laid out our shiny new, blue, aluminium, dinghy anchor. The children took us to the mayor's house to present our visitor permit and pay our dues. By obligation, Nicole wore a beautiful dress that was made for her by a Majuro dressmaker.

Dress code Aur (Taraipo in the distance)

Women must cover their shoulders at all times, a custom established by the London missionaries after the last World War. The dresses are beautiful, but not very practical when she had to swim back to Taraipo, because the wind had unexpectedly changed direction. Now on a lee shore, the waves were too big for two in the dinghy. The captain always being the last to leave the sinking ship. Once back aboard, it was anchors aweigh, and a quick passage to Tabal at the other end of the atoll before dusk and while the 'bommies' (uncharted coral heads) were still visible.

One day, after a social in the village, we returned to find our beautiful shiny blue dinghy anchor was missing. The kids took us to report this to the police. There are two policemen, one of them being the mayor, who delegated the job of solving the crime to his assistant. He sent us back to Tuggi to wait, as he was confident that the situation would be resolved quite quickly, and it was. The notion of property is that, if something is not secured, it can be shared. Hence, every door in the village has a flimsy padlock on it.

There was no malice; all good, clean fun. One fine day, Todd, a lanky American guy wearing specs, swam out to us. He clambered up the stern ladder and explained that he was a volunteer teacher from WorldTeach. He was spending a year living with a host family, and on their diet, most of which he could cope with, apart from 'day old fish with eyes for breakfast'. His house mother solved this problem by removing the head for him. He was a bit of a pied piper and often arrived with a gaggle of teenagers who enjoyed invading our privacy to see what we had and what they could beg from us. We were careful to hide away as much as possible and gifted them stuff they did not like too much, such as raisins or pasta, as we had no sweets on board. Adult visitors were more of a problem, as they would be looking for booze, which was illegal on the atolls, though we did happen to see an empty vodka bottle being used as a rolling pin. The kids let slip that they had seen on the way to school, one of their fathers, who had been tied to a tree for the night for being disorderly. We were always very generous with the teas and coffees and traded fishing hooks for coconuts. On Tabal, the ladies were very clear about the risk of being abused by the men if they wandered about unaccompanied. One night, one of them ran screaming through the village after being attacked by her husband, a respected member of the community. An experienced visiting teacher explained that not only do you wear a dress, you also need an underskirt, a pair of shorts and your knickers if you want to have a chance against the marauding men. No man had ever been convicted of domestic abuse, despite several cases being brought before the court in Majuro, which was presided over by male judges only.

Nicole spent many mornings ashore, sitting cross-legged in the shade, drinking coconut milk and nibbling fritters cooked on an open fire. She traded shell jewellery, which was very popular with the young girls. In exchange, the women invited Nicole to a party – aboard our boat. I was allowed to attend as an observer! They brought everything with them – the food, the ukulele, the singing and their infectious laughter. I was useful to the fulsome ladies in taxiing them in and out, carefully, one by one in Tuggii, our small dinghy.

Mischievous invaders

Ladies party, Aur

We were expected to pay our respects at a funeral that lasted for several days. The preparations required some of the young men to dive deep in the lagoon at night to catch a turtle big enough to feed the whole village. The poor beast died slowly, staked out in the sun belly up. Nicole found this a very upsetting spectacle. Hanake, who was head of the family that owns the small island where turtles are found, decided how many were taken each year and reminded us that 'every coconut belongs to somebody'. On the atolls, the meagre food chain is carefully managed. There is virtually no earth to grow things. Traditionally, the population survived on what could be taken from the sea plus coconuts and a few hardy vegetables. Unfortunately, the occupation of the islands by the Japanese and the Americans has upset the dietary balance of what was a healthy population a hundred years ago. They are now addicted to imported rice, sugar, white flour and red meat, and have become obese and were overwhelming the hospital in Majuro with diabetes patients. Tabal's 'doctor' had only one month's training before being designated to manage the small pharmacy that offers free medicines. He was the only person on Aur who listened to the inter-island radio net. His was the voice that we had come to know, and he took us under his wing when we arrived in Tabal.

Pure sailing

I was determined to go to Ailuk to witness the fastest outrigger sailing canoes still working daily. As usual, with the wind behind us, there was no chance of stopping. We had to find the entrance to the atoll and be careful not miss the second of the unmarked dark blue gaps through the reef. If only we had had a drone to assist us; they did not exist on 23 December 2010. Our bare legs were showered with warm, salty water as Taraipo skimmed through the channel in the coral reef. Once inside the atoll, the clouds blotted out the colour changes as we desperately tried to spot 'bommies' on the beat up to the south-east corner. It is always tricky trying to locate a patch of coral-free sand when arriving in an atoll, the worst situation being when the anchor chain jams around a coral head and becomes impossible to free; not good for the coral, not good for the chain (the galvanising gets scraped off and rusts), and not good for the nerves.

Wow. I was not disappointed with the intriguing line up of proas on the beach that have been improved by the use of some stainless fittings, rigging wire, and sails made from glued panels of plastic tarpaulin kept carefully in the shade when not sailing. The proas are super lightweight, excellent load carriers, fast and exciting to sail; not bad for your everyday workhorse carrying people and tons of coconuts from end to end of the lagoon. We arrived on Ailuk on Christmas Eve. We were invited to join festivities the next morning and were not sure what to expect. We were subjected to a morning of American 'ethnic' music played on the church organ (an electronic key board), and shuffle dancing in the church that had been cleared for the event. Each age group performed the same dance and sang to the same music. This was followed by a strange lunch al fresco in the hot sun. We, the visitors, sat cross-legged at the table and were served with a succession of dishes, chicken, fish and rice. We noticed that nobody else was having Christmas dinner. The villagers were standing around and watching us eat. It then dawned that they had not had any supplies since November. The irregular ferry from Majuro had just not come and we were eating their Christmas dinner. We left as much as possible on our plates without appearing impolite and, as we had guessed, the remainders were collected and put back in the shade. Another reason that supplies were low is that every family had given a gift to the church, a bag of rice, sugar or flour, all carefully stored in the minister's house.

On Boxing Day, the wind backed to the north-east, so we had to weigh anchor and retreat to Enijabro Island at the other end of the atoll for shelter from the waves that build quickly with a 15-mile fetch. The danger was that the anchor would drag and Taraipo would be wrecked on the coral. It is not a nice feeling when the rigging starts to hum in the night and the waves start breaking on the bows. We rowed ashore in Tuggi to experience an eerily deserted village, apart from a few pigs. The people had all gone to Ailuk for Christmas. The island chief, the only man with an outboard engine, came back from Ailuk to catch a pig to bolster food supplies. We had several visits from passing canoes on their way to harvest coconuts from the uninhabited islands nearby. The crews were as interested in our 'sailing canoe' as I was with theirs.

Sadly, we had to move on before I was able to persuade them to take me for a sail, something I yearned to do. I guess they thought that I might capsize the tippy proa and lose a whole day's harvest of coconuts. Imagine the scenario chasing a thousand coconuts floating downwind while the crew struggled to right the canoe! Proas always sail with the small hull (aka) to windward. This requires a tricky balancing act when they shunt to go in the opposite direction, releasing the tack of the sail and taking it to the opposite end of the main hull, and the steersman swaps ends with his paddle.

Off to collect coconuts in Ailuk

At dawn, as we were heading north towards Maloelap atoll when we suffered a big shift in wind direction and strength. We were no longer able to go towards our destination on starboard tack. I was fatigued after a long night on watch cajoling Taraipo into one more new and magical anchorage. We turned through the eye of the wind onto port tack and headed back towards the invisible north pass into Aur atoll. Having spied the dark-coloured water indicating the entrance, we reached into the lagoon and spent the rest of the morning threading our way through the coral heads. These uncharted 'bommies' rise vertically from the seabed and are impossible to spot without bright sunlight. Each time that a cloud crossed the sun, the 'bommies' disappeared! You only crash into a coral head once; the boat would be wrecked and, in the Marshall Islands, the islanders could sue you in court for damaging their habitat. Reassuringly, the 'new' engine could keep the boat edging forwards in a gale of wind. We were in control of our stressful situation. Nevertheless, that was the moment that I decided that I wanted to do something else with my life, after nine years of living in paradise.

By the time we had returned to our mooring in Majuro, Nicole had almost resigned herself to the fact that Taraipo was to be sold. Karen of the Mieco Beach Yacht Club said that we were crazy, so obviously happy and contented on our raft. "You will never sell the boat here in Majuro thousands of miles away from civilisation." It was true that we could have continued sailing around the world forever. However, after nine years afloat, pastures new beckoned. The world was moving on quickly and I wanted to reintegrate. Nicole did not want to disconnect from her family and friends as I had. So 'let's sell the boat and move to Switzerland'. We advertised the boat for sale on Scott Brown, the dependable catamaran broker's website. We had a 'bite' within 24 hours from a young Frenchman. He was an experienced sailor attracted by the fact that the boat was being sold complete and ready to sail to Australia, which would have been our next stopover had we kept Taraipo. Changing boat ownership and nationality while abroad was a bureaucratic, longwinded nightmare, and required diligence by the broker when the money was changing hands. Immigration had warned us that our six-month stopover was overdue. We flew out with hand luggage only and the exciting prospect for me of a new life in Switzerland, far from the sea. At the first stopover in Guam, I bought myself a pair of shoes and some socks, and threw away my smelly flip-flops. In Seoul, I bought a fleece; no need to replace my everlasting trousers, 'Rohan Bags'.

Two months later, Taraipo's new owner went back to France and left her in the care of an American sailor. The caretaker borrowed her and took her for an overnight joyride downwind to Mili Atoll. From there, he was evicted by the Mayor of Mili for not having a visitor permit from Majuro, the capital. Being an experienced sailor and, I guess, not relishing the prospect of sailing back upwind to Majuro, he sailed more than 1,000 miles to Guam, where he left Taraipo anchored in a deserted cove and disappeared ashore. She was impounded by the US coastguard and was happily sold locally.

Epilogue

Taraipo has a new owner and a family to sail her in Guam.

During autumn 2014, we concluded that the time was ripe to get married, quietly, secretly, in the City of Edinburgh. However, the word got out and, with a flourish of bagpipes, we had a party!

Four years after selling Taraipo, we spent two summers searching for a cottage to buy in the heart of Brittany, somewhere we could set up home and make our way. We found a large house overlooking the Atlantic Ocean, within budget, but I had an argument with the owner's wife while trying to do the deal. We drove away empty-handed, but with a deep feeling of relief!

We thought that we could live without a boat; we couldn't! We wanted to sail with our families and friends. I yearned to return to the waters of my apprentice years, cruising the creeks and bays of the coasts of South Devon, Cornwall and Brittany.

We now have a 10-metre keelboat, which is fun to sail, can sleep seven and is equipped (heating, fridge, AIS, chart plotter and engine) for sailing in the rough and tumble of the Western Approaches.

Glossary

Antifouling – paint to prevent biofouling below the waterline

Bilge – the inside bottom of a boat

Boom – metal spar on the lower edge of the mainsail

Bommie – coral head

Bowline – knot to make a non-slip loop

Capsize – to accidently overturn a boat

Cleat – fastening to hold a rope

Chandlery – boat equipment shop

Close-hauled – sailing close to the wind

Dinghy – small boat

Dress overall – to hoist all code flags (alphabet+)

Draft – depth below waterline

EPIRB – emergency position-indicating radio beacon

Fender – bumper to prevent damage

Flake – lay chain in a 'zigzag'; pile rope in a heap

Gale – strong wind 39–54 miles/hour

GPS – satellite positioning instrument

Halliard – rope to hoist a sail

Haul-out – to lift a boat ashore

Jackstay – webbing safety line to clip onto

Jib – front sail

Jury rig – makeshift repair

Kedge – spare anchor

Keelboat – mono-hulled boat with a ballast keel

Occulting light – shows 'black flashes'!

Pan-Pan – urgent safety message

Port – left side facing forward

Reaching – to sail with the wind at right angles to the boat

Reef – to reduce sail size

RIB – rigid-hulled inflatable boat

Rudder – lower part of the steering apparatus

Saltire – the Scottish flag

Sheet – a rope to control a sail

Spinnaker – a downwind sail

Spring – a mooring rope

Starboard – right side facing forward

Stem – front end of a hull

Storm – Strong wind 55–63 miles/hour + rain and lightning

Tiller – steering arm

Topping lift – rope to hold up the boom

VHF – short-range radio transmitter

Warp – rope for tying up a boat

Winch – device to pull in the sheets

Windlass – device to pull up the anchor chain

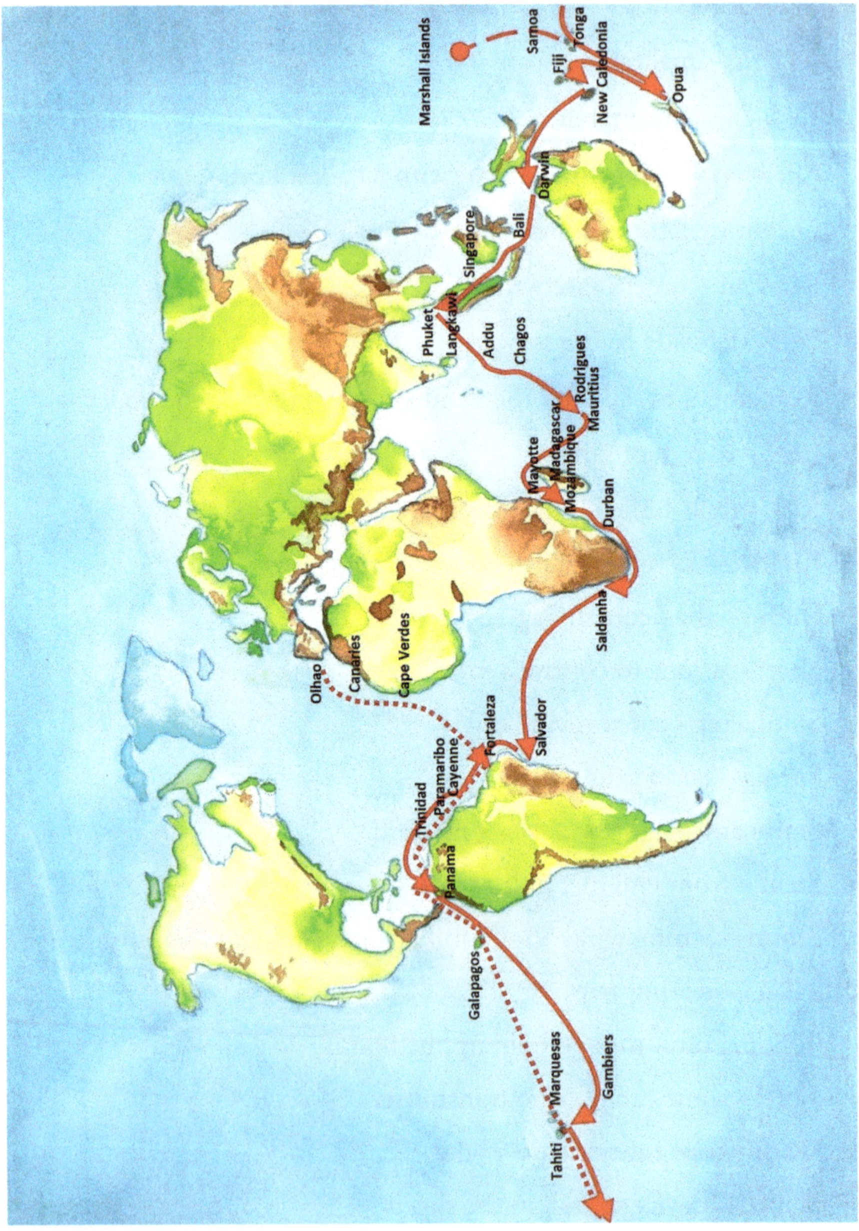
Marshall Islands
Samoa
Tonga
Fiji
New Caledonia
Opua
Darwin
Bali
Singapore
Langkawi
Phuket
Addu
Chagos
Rodrigues
Mauritius
Mayotte
Madagascar
Mozambique
Durban
Saldanha
Canaries
Cape Verdes
Olhao
Fortaleza
Salvador
Cayenne
Paramaribo
Trinidad
Panama
Galapagos
Marquesas
Gambiers
Tahiti

About the author

John Jameson, of the Irish whiskey family, was born in 1941 in Cheshire. At the age of 19, he went to sea as crew on the ship Lochinvar and continued delivering yachts. He became a sailing instructor and skipper at the Island Cruising Club, and a manager at the Glenans sailing school in France. Later, he moved back to the UK, where he worked as Derbyshire Education County Sailing Master, and was RYA Scottish National Sailing Coach for 17 years. In 2000 he set off to sail around the world aboard a Polynesian Catamaran.

Lightning Source UK Ltd.
Milton Keynes UK
UKHW051156300720
367370UK00011B/105

9 781839 751240